Math Mammoth
Grade 8 Tests and
Cumulative Revisions

International Version

Includes consumable student copies of:

- Chapter Tests
- End-of-year Test
- Cumulative Revisions

By Maria Miller

Contents

Grade 8, Chapter 1

End-of-Chapter Test

Instructions to the student:

A calculator is *not* allowed for questions 1-5. A calculator *is* allowed for questions 6-8.
Answer each question in the space provided.

Instructions to the teacher:

A calculator is *not* allowed for questions 1-5. A calculator *is* allowed for questions 6-8.

My suggestion for grading the chapter 1 test is below. The total is 33 points. Divide the student's score by the total of 33 to get a decimal number, and change that decimal to percent to get the student's percentage score.

Question #	Max. points	Student score
1	8 points	
2	8 points	
3	3 points	
4	3 points	
5a	1 point	
5b	1 point	

Question #	Max. points	Student score
6	2 points	
7	2 points	
8	3 points	
9	2 points	
TOTAL	**33 points**	/ 33

Chapter 1 Test

Do not use a calculator for the problems on this page.

1. Find the value of each expression, without a calculator.

a.	b.	c.	d.
$(-9)^2 =$	$5 \cdot 2^{-3} =$	$5^8 \cdot 5^3 \cdot 5^{-9} =$	$\left(\dfrac{1}{4}\right)^3 =$
$-9^2 =$	$(5 \cdot 2)^3 =$	$(4 \cdot 3)^{-2} =$	$\dfrac{5^7}{5^9} =$

2. Write an equivalent expression, using the exponent laws, and without leaving negative exponents.

a. $3v^5 \cdot 2^3 \cdot v^3 =$	**b.** $-2b^5 a^7 \cdot 5b^3 a^8 =$	**c.** $(5x)^2 =$	**d.** $(2x)^{-3}$
e. $2s^{-2} =$	**f.** $(s^{-2})^3 =$	**g.** $\left(\dfrac{-x^2}{2}\right)^5 =$	**h.** $\dfrac{25x^9}{15x^4} =$

3. Find the expressions that have the value 3^8.

a. $3^2 \cdot 3^3 \cdot 3^3$	**b.** $\dfrac{3^{16}}{3^2}$	**c.** 24	**d.** $\dfrac{3^{11}}{3^3}$	**e.** $\dfrac{3^7}{1/3}$	**f.** 3^{2^4}

4. Order the numbers from the smallest to the largest.

$2 \cdot 10^8$ $0.4 \cdot 10^6$ $5.2 \cdot 10^7$ $9 \cdot 10^5$ $64 \cdot 10^6$

5. **a.** How many times bigger is $7 \cdot 10^{-9}$ than 10^{-18}?

b. How many times bigger is $2 \cdot 10^6$ than $8 \cdot 10^{-5}$?

6. A (rectangular) driveway is measured to be 9.91 m long and 3.24 m wide. Calculate its area, giving the result to a reasonable accuracy.

7. Calculate an estimation for the total cost of providing 12 000 students with a meal costing $2.65 each, five days a week, 36 weeks a year. Assume that 12 000 is accurate to three significant digits, and treat five and 36 as perfectly accurate.

8. The speed of sound in salt water is about 1500 m/s. Calculate the distance sound travels in a year, in salt water. Give your answer in kilometres, in scientific notation, and with two significant digits.

9. The mass of one copper atom is about $1.055 \cdot 10^{-22}$ grams. How many atoms are in 10 grams of copper? Give your answer in scientific notation.

Grade 8, Chapter 2

End-of-Chapter Test

Instructions to the student:

You may use a calculator. Answer each question in the space provided.

Instructions to the teacher:

The student is allowed to use a basic calculator in this test (not a graphing calculator).

My suggestion for grading the chapter 2 test is below. In the questions that ask a student to explain or justify their reasoning, the reasoning portion is worth 2 points and the numerical answer is worth 1 point. So, if the student only writes down a numerical answer, without explaining the reasoning, the student gets one point out of three.

The total is 30 points. Divide the student's score by the total of 30 to get a decimal number, and change that decimal to percent to get the student's percentage score.

Question #	Max. points	Student score
1	2 points	
2	3 points	
3	2 points	
4	2 points	
5a	1 point	
5b	1 point	
5c	1 point	
5d	1 point	

Question #	Max. points	Student score
6	3 points	
7	3 points	
8	3 points	
9a	1 point	
9b	2 points	
10	2 points	
11	3 points	
TOTAL	**30 points**	/ 30

Chapter 2 Test

1. Trapezium ABCD is first rotated 90 degrees counterclockwise around the origin, and then reflected in the horizontal line $y = 1$.

   Draw the image of the trapezium after these transformations.

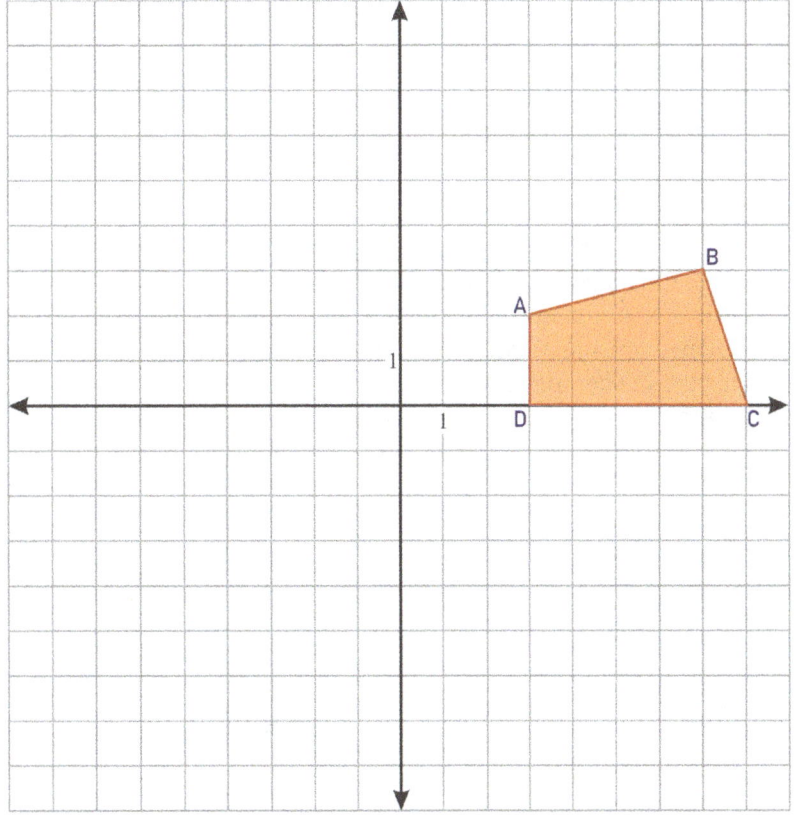

2. Show that the two triangle shapes are similar by describing a sequence of transformations that maps the larger arrow to the smaller arrow.

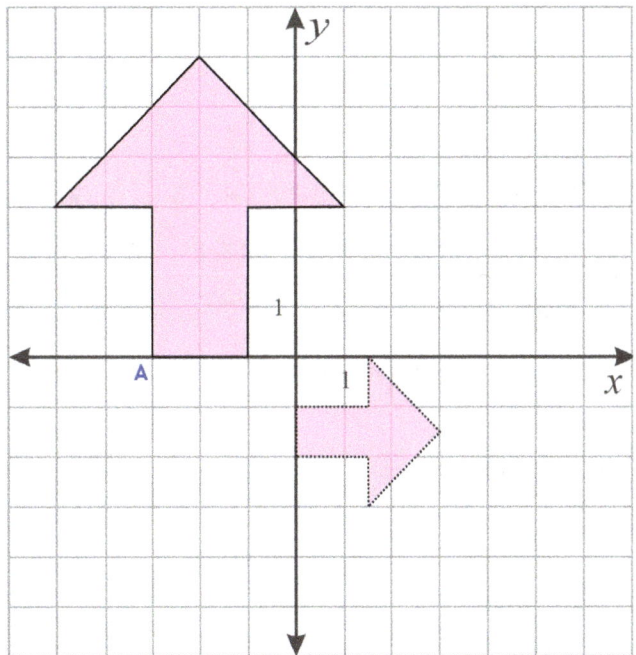

3. A triangle was first reflected in the *y*-axis, then translated two units down. Now its vertices are at points (−4, −2), (−3, 0), and (−1, −2). What were the coordinates of its vertices before these transformations?

4. A rectangle with vertices at (9, −3), (9, 0), (3, −3) and (3, 0) is dilated with origin as the centre of the dilation, and with a scaling factor of 1/3. Then it is translated one unit to the left and one unit down. What are the coordinates of the vertices of the resulting rectangle?

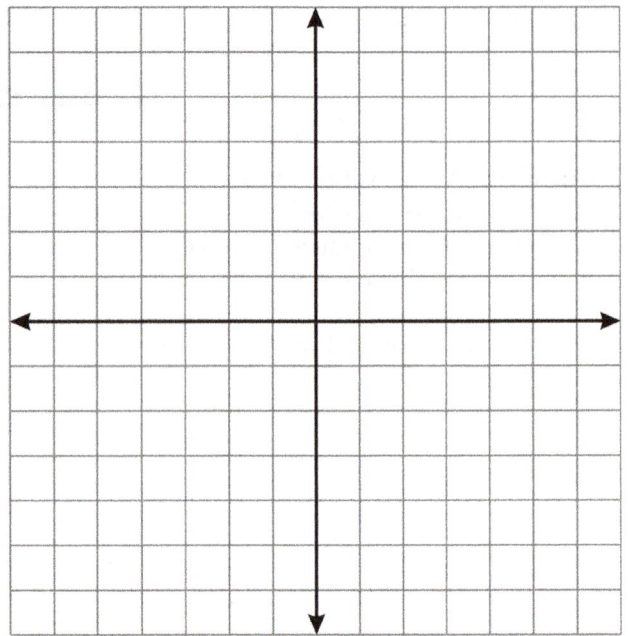

5. Quadrilateral ABCD undergoes a dilation (from 1 to 2) and then a certain congruent transformation, from 2 to 3.

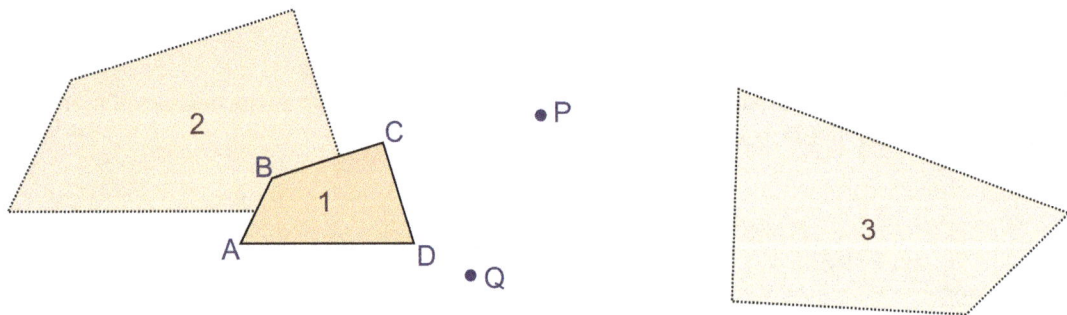

   a. Name the congruent transformation.

   b. Mark as A' the image of point A under the dilation.

   c. Mark as A" the image of point A' under the other transformation.

   d. Which attributes of the quadrilateral are preserved in this sequence of transformations? Tick all that apply.

   (i) Perimeter

   (ii) Area

   (iii) Position

   (iv) Angle at A

   (v) Angle sum

6. Figure ABCD is a parallelogram.
   What is the value of x?

   Explain your reasoning.

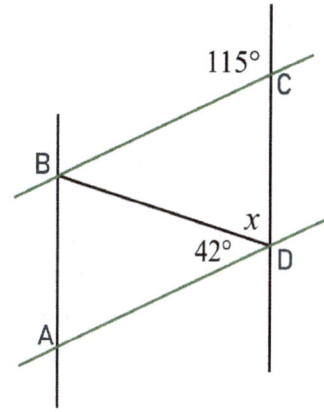

7. Find the measure of angle x.
   Explain your reasoning.

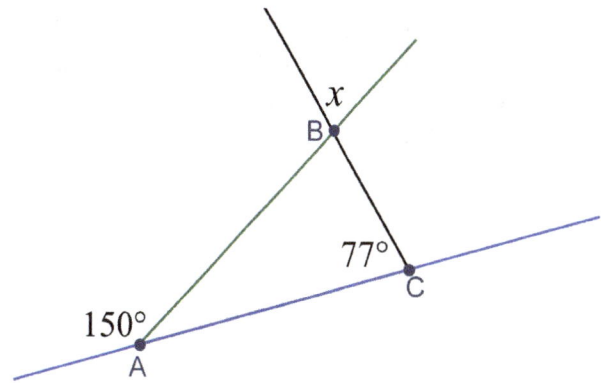

8. Are the two triangles similar?
   Explain how you know.

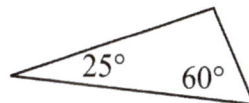

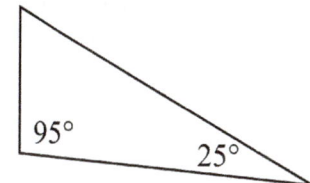

9. **a.** Find the volume of a sphere with a radius of 4.0 cm. Give your answer to a reasonable accuracy, based on significant digits.

   b. Find the volume of a cylinder, when its bottom face is a circle with a 4.0-cm radius, and its height is twice the diameter of that circle. Give your answer to a reasonable accuracy, based on significant digits.

10. What fraction is the volume of this cone of the volume of a cylinder it just fits into? In other words, the cylinder and the cone have the same size circle as their base, and have the same height.

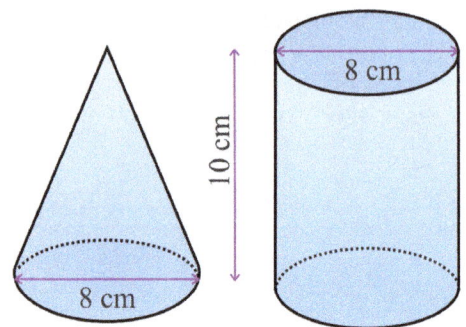

11. Kayla says, "These four limes will give me a half a cup of lime juice." Should you believe Kayla? The limes are spherical, with a diameter of 6 cm, one cup is 240 ml, and 1 ml = 1 cm^3. Justify your reasoning.

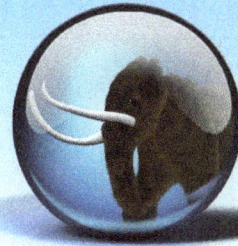

Grade 8, Chapter 3

End-of-Chapter Test

Instructions to the student:

You may use a basic calculator. Answer each question in the space provided.

Instructions to the teacher:

The student is allowed to use a basic calculator in this test (not a graphing calculator).

My suggestion for grading the chapter 3 test is below. The total is 24 points. Divide the student's score by the total of 24 to get a decimal number, and change that decimal to percent to get the student's percentage score.

Question #	Max. points	Student score
1	8 points	
2	2 points	
3	3 points	
4	3 points	
5a	1 point	
5b	1 point	

Question #	Max. points	Student score
6a	1 point	
6b	1 point	
7	4 points	
TOTAL	**24 points**	/ 24

Chapter 3 Test

1. Solve.

a. $6 - 3y - 2 + 8y = 4y + 1 - 9y - 5$	**b.** $12 - (x - 2) = 10 - 2x$
c. $4x + 20 - x = 2(x - 5) - 6x$	**d.** $-35 - 3(x - 4) = 10x + 40 - x + 13x$

2. Ohm's law concerning electric circuits states that $C = \dfrac{V}{R}$, where C is the current in the circuit,

V is the voltage in the circuit, and R is the resistance in the circuit.
Solve this formula for R.

3. Lucas says, "In six years, I will be two-thirds as old as my mom, who is now 66." How old is Lucas?

4. A storekeeper sells a chainsaw for $288. By how much should he increase the price, so that when he later on has a 20% off sale, the sale price will be $249?

5. **a.** How many solutions does this equation have?

$$8y + 4 = 4(3 - 2y)$$

   b. Modify the equation so that it has an infinite number of solutions.

6. **a.** How many solutions does this equation have?

$$-2(9y - 4) + 8y = 8 - 10y$$

   b. Modify the equation so that it has no solutions.

7. Solve.

a. $\dfrac{3x - 5}{4} - 1 = 2x$	**b.** $\dfrac{y - 2}{3} = 3y + \dfrac{5 - y}{2}$

Grade 8, Chapter 4

End-of-Chapter Test

Instructions to the student:

Answer each question in the space provided. Do not use a calculator.

Instructions to the teacher:

The student is *not* allowed to use a calculator.

My suggestion for grading the chapter 4 test is below. The total is 32 points. Divide the student's score by the total of 32 to get a decimal number, and change that decimal to percent to get the student's percentage score.

Question #	Max. points	Student score
1a	2 points	
1b	2 points	
2a	2 points	
2b	1 point	
2c	3 points	
3	4 points	
4a	1 point	
4b	1 point	
4c	1 point	
4d	1 point	

Question #	Max. points	Student score
5	5 points	
6a	2 points	
6b	2 points	
6c	1 point	
6d	2 points	
6e	1 point	
6f	1 point	
TOTAL	**32 points**	/ 32

Chapter 4 Test

1. **a.** Change some thing(s) in this table so it represents a function.

   b. If you reversed the inputs and the outputs, would the resulting relationship still be a function? Explain.

Input	Output
Name	*work hours*
Sally	5
Susan	6
Jane	10
Joe	?
Jane	8
Harry	0
Mike	7
Rob	9
John	5

2. Three functions are represented below.

   a. Which one has the largest initial value? Also, state that value.

   b. Which one(s) are linear functions?

   c. Find the rate of change for each function in the *x*-interval [7, 10].

   Function 1:   Function 2:   Function 3:

Function 1:

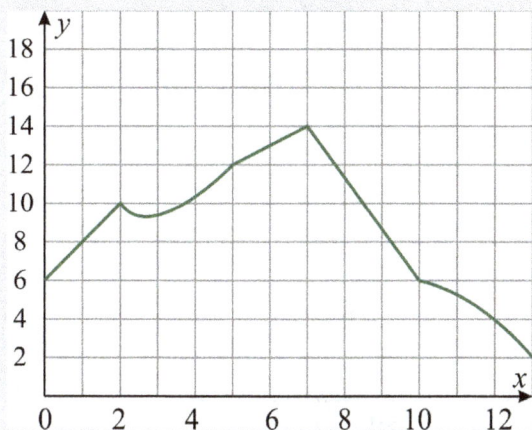

Function 2:

$y = 15 - 1.2x$

Function 3:

x	*y*
0	16
2	14
4	12
6	11
7	10
8	8
10	5
12	2

3. Describe this function by intervals where it is increasing, decreasing, or constant. Include also whether it is linear or nonlinear in those intervals.

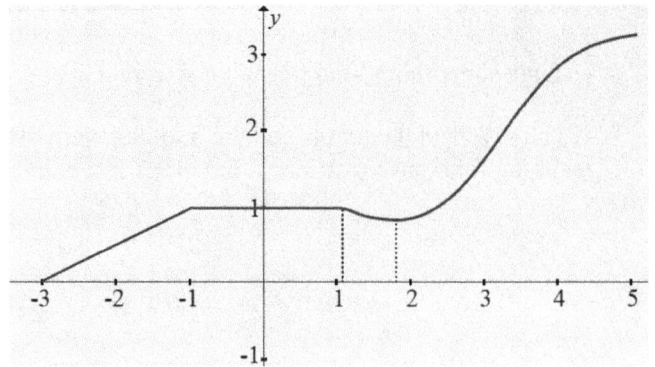

4. Leo is saving money at a steady rate. The equation $M = 150w + 200$ describes the number of dollars he has, as a function of time, where w is the number of weeks since January 1.

   a. Is this a linear function?

   b. What is the rate of change? Include the units.

   c. What is the initial value?

   d. After how many whole weeks will Leo have at least $1200?

5. **a.** Sketch a plot of a function that depicts the distance Marsha has walked that matches the story below.

   For the first 10 minutes, Marsha walks at a fairly brisk and steady pace of 6 km/h. Then she stops for 10 minutes to talk with a friend at a park. Next, she slowly walks a distance of 1 km over the next 15 minutes. She stops again for five minutes. Lastly, she picks up speed and walks at a pace of 8 km/h for 10 minutes.

   b. What is the total distance Marsha walked?

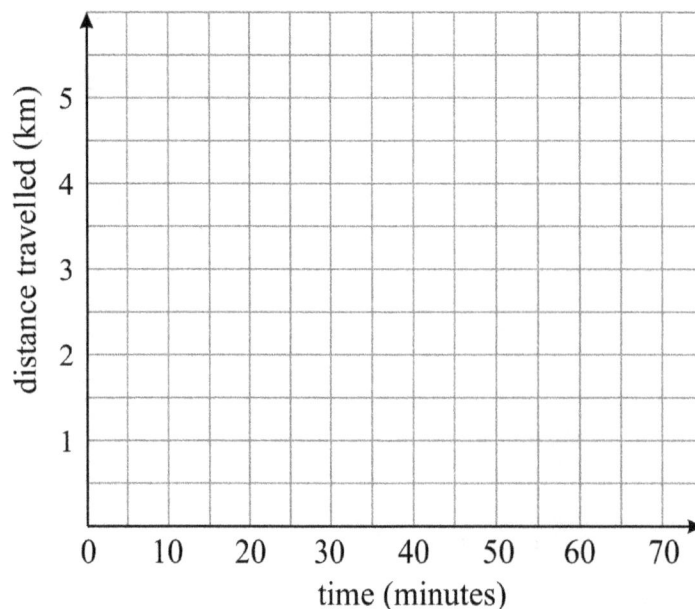

6. Janet starts filling a tub of water with a hose. The volume of water in the tub is a function of time *(t)*.

t (minutes)	0	1	2	3	4
V (litres)	48	64	80	96	112

a. What is the rate of change?

What does it mean in this situation?

b. What is the initial value?

What does it mean in this situation?

c. Write an equation that gives you the volume of water as a function of time.

d. Plot the equation you wrote in (c).

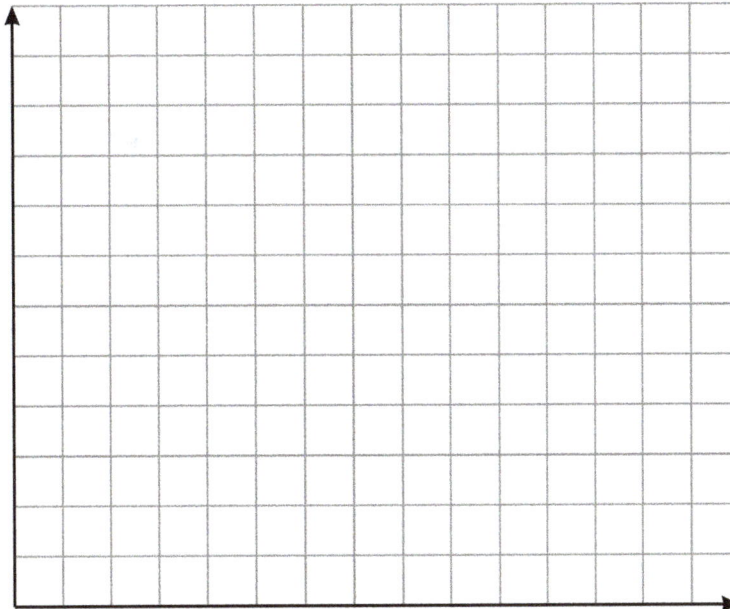

e. How much water is in the tub after 10 minutes?

f. When will the tub have 128 litres of water in it?

Grade 8, Chapter 5

End-of-Chapter Test

Instructions to the student:

Answer each question in the space provided. A basic calculator *is* allowed (not a graphing calculator).

Instructions to the teacher:

The student *is* allowed to use a <u>basic</u> calculator. A graphing calculator is not allowed.

My suggestion for grading the chapter 5 test is below. The total is 35 points. Divide the student's score by the total of 35 to get a decimal number, and change that decimal to percent to get the student's percentage score.

Question #	Max. points	Student score
1a	1 point	
1b	1 point	
1c	2 points	
1d	2 points	
1e	1 point	
2a	2 points	
2b	2 points	
2c	2 points	
2d	2 points	

Question #	Max. points	Student score
3a	1 point	
3b	2 points	
3c	1 point	
4a	2 points	
4b	2 points	
5	3 points	
6a	4 points	
6b	2 points	
7	3 points	
TOTAL	**35 points**	/ 35

Chapter 5 Test

1. Liam is buying paint for his house. He is considering two different types of paint, Paint 1 and Paint 2. The equation A = 12p gives you the area (A), in square metres, that p litres of Paint 1 covers. The graph shows how much area Paint 2 covers.

Paint 1: A = 12p   **Paint 2:**

area
(m^2)

```
900
800
700
600
500
400
300
200
100
        10    20    30    40    paint (litres)
```

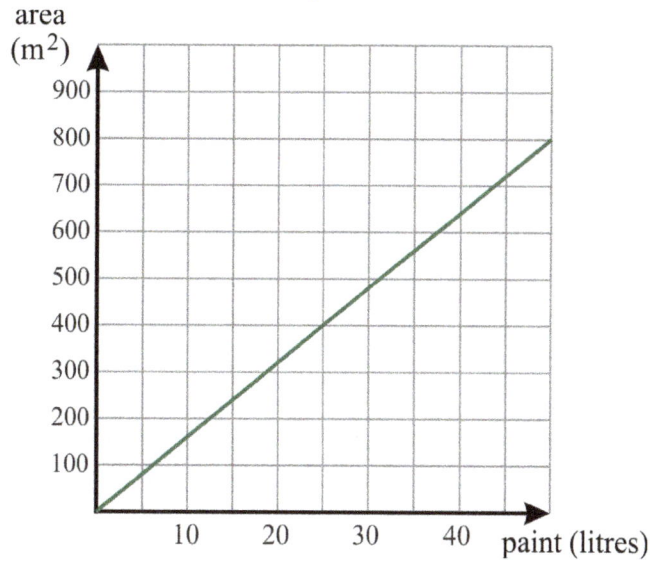

a. What is the slope of the line depicting the coverage of Paint 2?

b. Which paint covers more area per litre?

c. The walls of a bedroom are 30 m^2. Calculate how much of each kind of paint would be needed to paint the room.

d. Draw a line depicting the area that Paint 1 covers, in the same grid as for Paint 2.

e. How much more does 40 litres of the one paint cover than 40 litres of the other?

2. Find the equation of each line, in slope-intercept form:

    a. has slope 3/4 and passes through (0, −8)

    b. has slope −5/2 and passes through (−2, 3)

    c. is vertical and passes through (−11, 4)

    d. is parallel to $y = 6x - 9$ and crosses the line $x = 4$ at $y = 2$.

3. Henry has been helping his obese cat to lose weight. He used to weigh 7.6 kg, but started losing weight in a steady manner so that a month later, he weighed 7.2 kg and two months later 6.8 kg.

    a. Consider the cat's weight as a function of time, and write an equation for it.

    b. Graph your equation. Design the scaling for the vertical axis so that the point (0, 7.6 kg) fits in it.

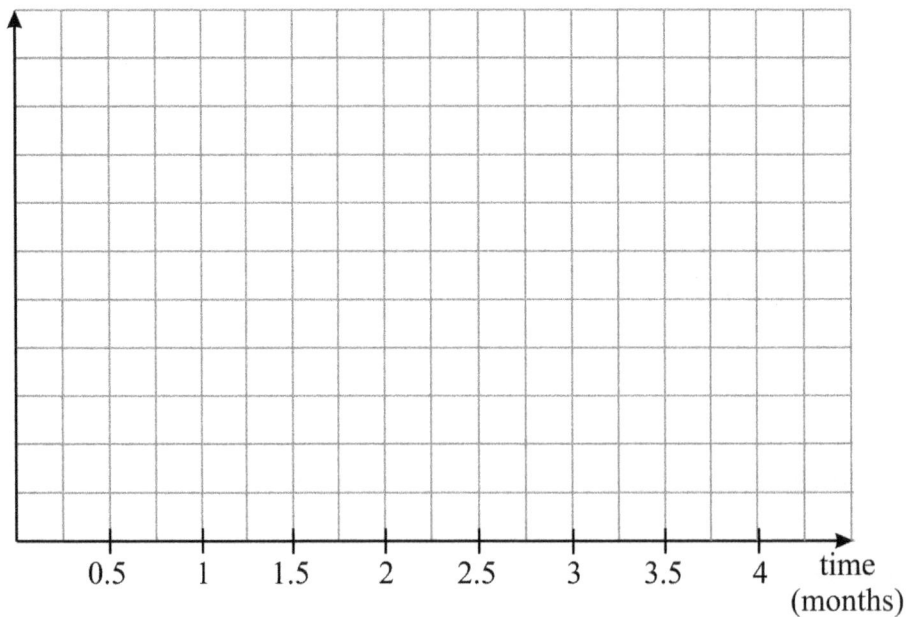

                                       0.5   1   1.5   2   2.5   3   3.5   4   time
   (months)

    c. In how many months will the cat weigh 5 kg if this continues?

4. Find the equations for the lines. Notice the scaling.

   a.

   b.

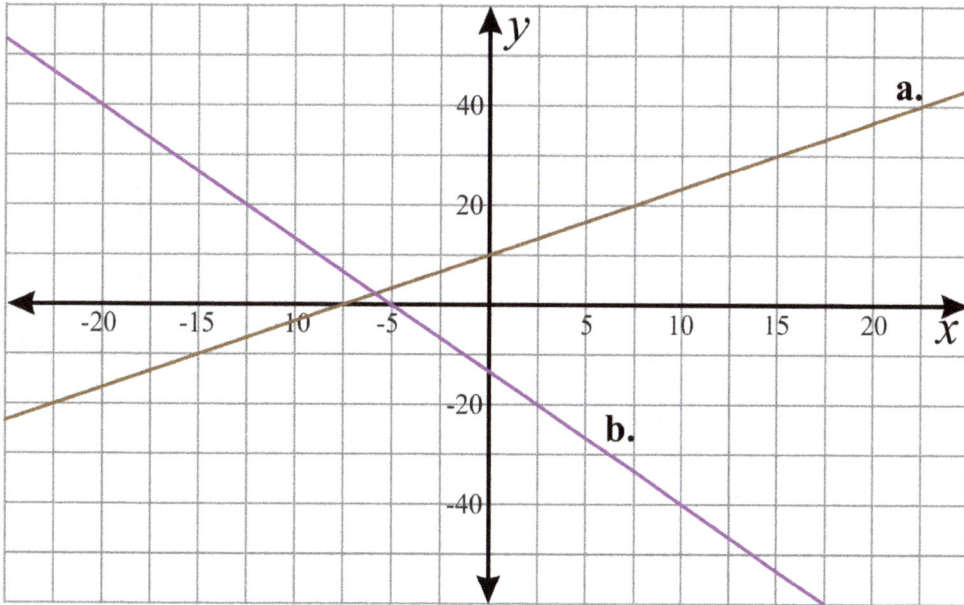

5. Write the equation $9x - 3y = 6$ in the slope-intercept form, and graph it.

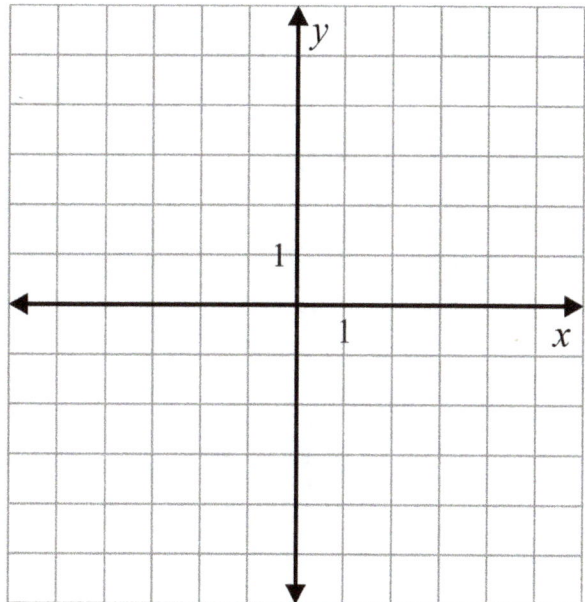

6. Line L passes through $(2, -1)$ and $(-3, 6)$. Line M is perpendicular to Line L, and passes through $(-2, -3)$.

    a. Find the equations of both lines, in slope-intercept form.

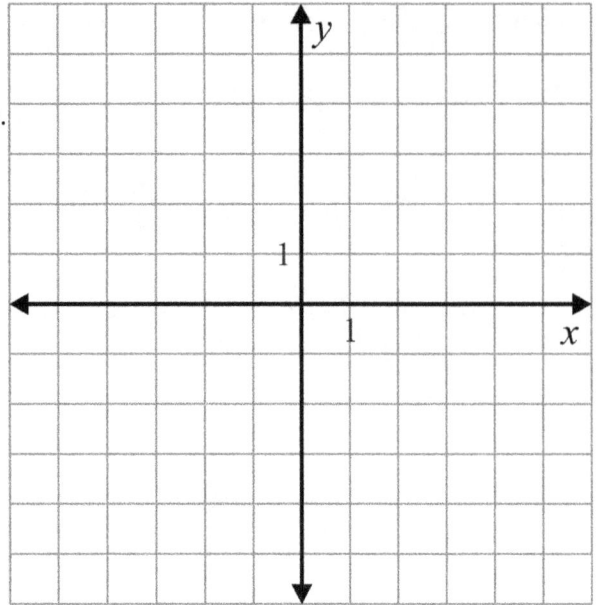

    b. Write the equations also in the standard form.

7. Find s so that the point $(6, s)$ will fall on the same line as the points $(1, -1)$ and $(10, -28)$.

Grade 8, Chapter 6

End-of-Chapter Test

Instructions to the student:

Answer each question in the space provided. A calculator is *not* allowed for the first page of the test (questions 1-5). A calculator *is* allowed for the second page (questions 6-8). Please give the first page of the test to your teacher before receiving the calculator.

Instructions to the teacher:

The student is *not* allowed to use a calculator for the first page of the test (questions 1-5). A calculator *is* allowed for the second page (questions 6-8). Collect the first page of the test before giving the calculator to the student.

My suggestion for grading the chapter 6 test is below. The total is 31 points. Divide the student's score by the total of 31 to get a decimal number, and change that decimal to percent to get the student's percentage score.

Question #	Max. points	Student score
1	4 points	
2	3 points	
3	6 points	
4a	2 points	
4b	2 points	
4c	1 point	

Question #	Max. points	Student score
5	3 points	
6	3 points	
7	3 points	
8	4 points	
TOTAL	**31 points**	/ 31

Chapter 6 Test

1. Find the correct statements.

   a. 0.141414 is rational because its decimal expansion repeats.

   b. $\sqrt{111}$ is irrational because it is a square root of a number that is not a perfect square.

   c. 2π is irrational because π is irrational, and an irrational number multiplied by a rational number is irrational.

   d. $\dfrac{\sqrt{64}}{3}$ is rational because it is a ratio of two whole numbers.

2. Write the two consecutive integers between which the square or cube root lies.

a. _____ < $\sqrt{44}$ <_____	**b.** _____ < $-\sqrt{5}$ <_____	**c.** _____ < $\sqrt[3]{21}$ <_____

3. Plot the following numbers *approximately* on the number line. Do not use a calculator.

   $-\sqrt{25}$ $\sqrt{80}-5$ $\sqrt[3]{27}$ $-2\sqrt{2}$ π $\sqrt{35}/2$

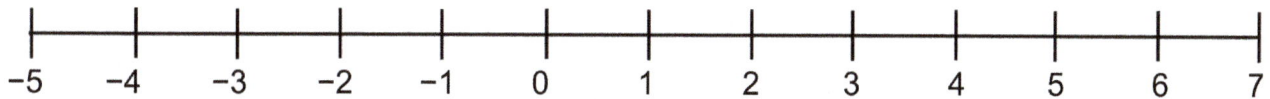

4. Solve. Give the final answers in exact form.

a. $x^2 = 37$	**b.** $5a^2 = 90$	**c.** $w^3 = 125$

5. Write the repeating decimal $0.\overline{23}$ as a fraction.

6. The two legs of a right triangle are $\sqrt{21}$ and $\sqrt{15}$. How long is the hypotenuse?

7. Find the distance between $(-2, 5)$ and $(10, 8)$. Give your answer rounded to two decimals.

8. Elizabeth runs around this track for morning exercise. Find the total distance she runs if she does three loops around this track.

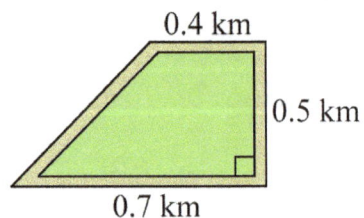

0.4 km

0.5 km

0.7 km

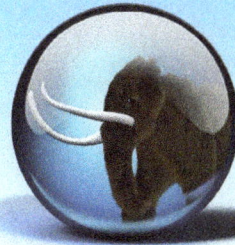

Grade 8, Chapter 7

End-of-Chapter Test

Instructions to the student:

Answer each question in the space provided. A basic calculator *is* allowed (not a graphing calculator).

Instructions to the teacher:

The student *is* allowed to use a <u>basic</u> calculator. A graphing calculator is not allowed.

My suggestion for grading the chapter 7 test is below. The total is 27 points. Divide the student's score by the total of 27 to get a decimal number, and change that decimal to percent to get the student's percentage score.

Question #	Max. points	Student score
1	3 points	
2a	1 point	
2b	1 point	
2c	1 point	
3a	3 points	
3b	3 points	

Question #	Max. points	Student score
4	2 points	
5	3 points	
6	3 points	
7	3 points	
8	4 points	
TOTAL	**27 points**	/ 27

Chapter 7 Test

1. Tell how many solutions each system of equations has. (You do not have to find the solution(s).)

    a. $\begin{cases} -y + 2x = 3 \\ 3y + 7x = -20 \end{cases}$ **b.** $\begin{cases} 3y - 4x = 6 \\ -6y + 8x = -12 \end{cases}$ **c.** $\begin{cases} 4x - 11y = -1/2 \\ -11y = 6 - 4x \end{cases}$

2. **a.** Give a value to s in such a manner that the system below has no solutions.

    $\begin{cases} y = 3x + 1 \\ y = sx - 2 \end{cases}$

    b. Graph the lines.

    c. Explain how the graph shows that there are no solutions.

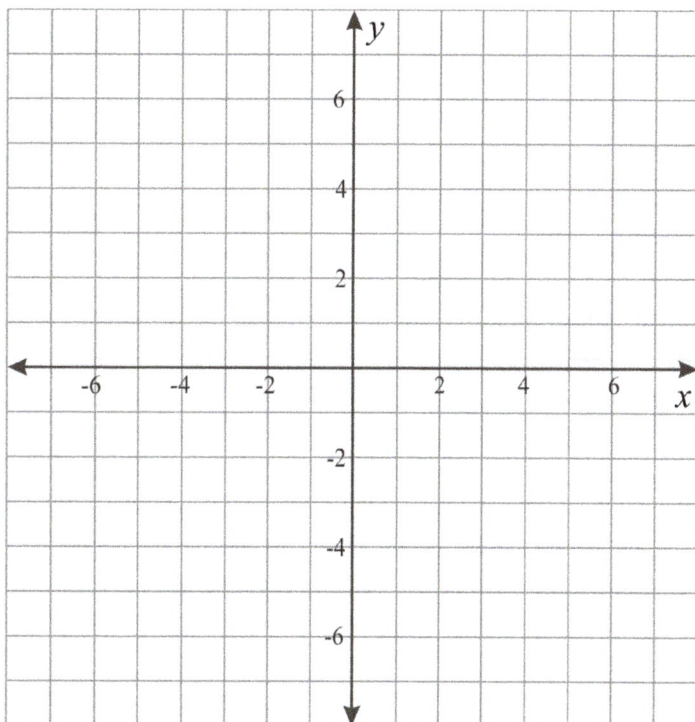

3. Solve each system of equations.

    a. $\begin{cases} x = 10 - y \\ 5y = 7(x - 2) \end{cases}$ **b.** $\begin{cases} 3x - y = 14 \\ 6x + 3y = -12 \end{cases}$

4. Solve the system of equations using any method. Also, graph the lines.

$$\begin{cases} x + 3y = -2 \\ 2x - 5y = -15 \end{cases}$$

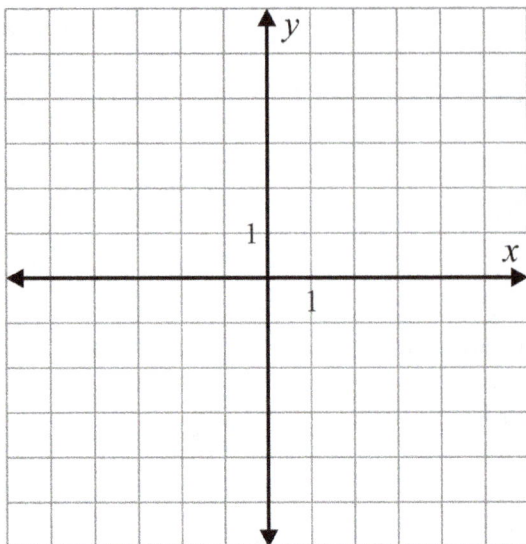

Solution: _____

5. Solve the system of equations and give the solution as decimals, rounded to three decimal digits.

$$\begin{cases} x + 6y = -1 \\ y = 2.2x + 0.8 \end{cases}$$

6. Ann and Adam talked about their ages. Ann said, "In three years, I will be double your age." Adam said, "And seven years ago, you were 2.5 times my age." Find their ages now.

7. A restaurant has two kinds of tables: tables that seat 4 people, and tables that seat 6. If the restaurant has a total of 28 tables, and those can seat 132 people, how many tables that seat 6 do they have?

8. Train 1 leaves Station A, heading for Station B, a distance of 70 km, at 3 PM and travels with a constant speed of 90 km/h. Train 2 leaves Station B, heading for Station A and travels at the constant speed of 105 km/h, at the same time.

At what time do they meet?

What distance has the second train travelled by that time?

	distance	velocity	time
Train 1			
Train 2			

Grade 8, Chapter 8

End-of-Chapter Test

Instructions to the student:

Answer each question in the space provided. A basic calculator *is* allowed (not a graphing calculator).

Instructions to the teacher:

The student *is* allowed to use a <u>basic</u> calculator. A graphing calculator is not allowed.

My suggestion for grading the chapter 8 test is below. The total is 25 points. Divide the student's score by the total of 25 to get a decimal number, and change that decimal to percent to get the student's percentage score.

Question #	Max. points	Student score
1a	2 points	
1b	1 point	
1c	1 point	
1d	2 points	
1e	1 point	
1f	2 points	
1g	2 points	

Question #	Max. points	Student score
2a	2 points	
2b	1 point	
2c	1 point	
2d	1 point	
3a	3 points	
3b	1 point	
3c	1 point	
4a	1 point	
4b	3 points	
TOTAL	**25 points**	/ 25

Chapter 8 Test

1. James started a training program to improve his 5-kilometre running time. The plot below shows the number of days he has been doing the program versus his running time in minutes.

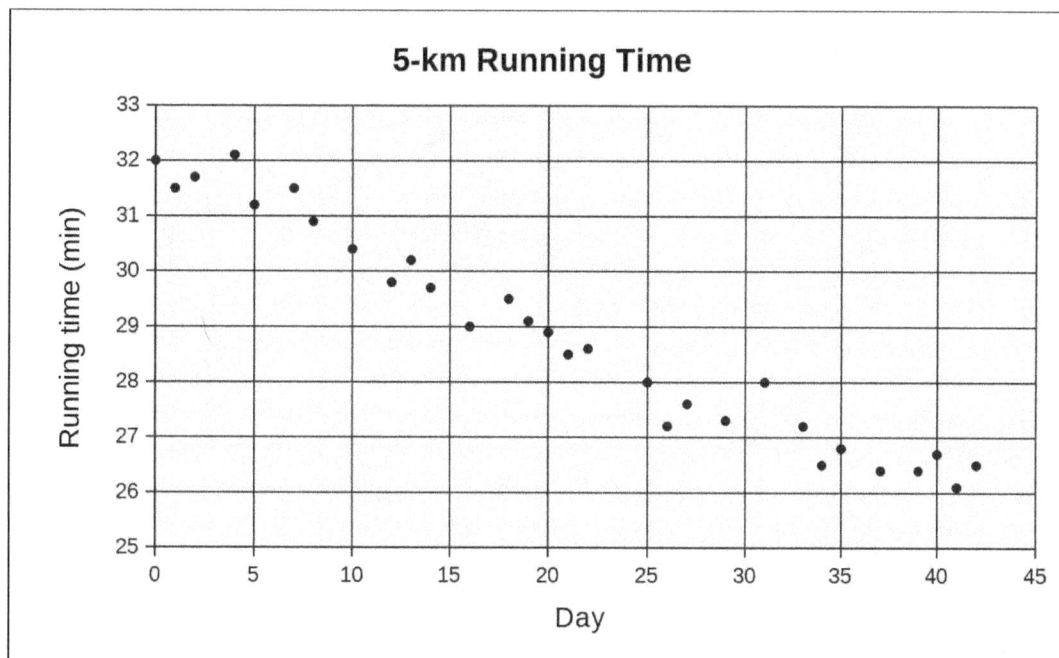

5-km Running Time

a. Is there a relationship or an association between the two variables? If so, what kind?

b. Which of the following points would be considered an outlier, if it was added to the plot?

(24, 28.3), (24, 29), (24, 31.3)

c. Draw a trend line to fit the data.

d. Find the (approximate) equation of your line.

e. What running time does your equation predict for day 24?

f. What does the slope of your equation signify in this context?

g. What does the *y*-intercept of your equation signify in this context?

2. The scatter plot below shows the diameter of pizzas from various pizzerias versus their price.

a. The equation of the trend line is $y = 0.528x - 5.95$.
What does the slope of this equation signify in this context?

b. Can we extrapolate and use the equation to predict the cost of, say, a pizza with 3-cm diameter? Explain why or why not.

c. Using the equation, predict what the price would be for a pizza with a 40-cm diameter.

d. Using the equation, predict the diameter of a pizza that costs $15.

3. The two-way table shows how many students in a certain high school didn't take foreign language, or took Spanish, French, German, or some other foreign language.

	None	Spanish	French	German	Other	Totals
Male	117	20	4	3	1	**145**
Female	97	40	9	3	2	**151**
Totals	**214**	**60**	**13**	**6**	**3**	**296**

a. Is there an association between the variables? Justify your answer.

b. If you choose randomly a student who took French, what is the chance they're female? Give your answer to the nearest percent.

c. If you choose randomly a male student, what is the chance they didn't take any foreign language? Give your answer to the nearest percent.

4. **a.** Find the two-way table with **no association** between the variables.

b. Explain how you found it.

Daily hours on social media	Exercise frequency per week				
	Never	**Once**	**Twice**	**Thrice**	**4x or more**
1	4	5	10	13	11
2	9	10	7	7	5
3-4	19	10	9	3	2
5-6	24	13	5	2	0

	Ate fish in the previous week?	
	Yes	**No**
Australia	23	20
Japan	87	2
Portugal	73	14
United States	56	47
Brazil	16	68

	Has two or more siblings	
	Yes	**No**
14-15 years	13	28
16-17 years	6	12
18-19 years	14	27

	Dropped out of school	
	Yes	**No**
high school	98	673
vocational school	93	367

End-of-Year Test — Grade 8

This test is quite long, because it contains questions on all of the major topics covered in *Math Mammoth Grade 8*. Its main purpose is to be a diagnostic test: to find out what the student knows and does not know of the topics covered in the curriculum. Since the curriculum follows the Common Core Standards for 8th grade, the test questions cover most of those standards.

Since the test is fairly long, I don't recommend that you have the student do it in one sitting. Feel free to break it into 2-5 parts and administer them on consecutive days, or perhaps in a morning/evening/morning/evening. Use your judgment.

<u>Important:</u> **A calculator is *not* allowed in the first two sections: Exponents and Scientific Notation, and Irrational Numbers. A basic calculator (not a graphing calculator) *is* allowed for the rest of the test.** The questions where calculator usage is most appropriate show a little calculator picture on the right.

The test is evaluating the student's ability in the following content areas:

- exponent laws
- scientific notation, including in calculations
- irrational numbers
- simple equations involving square or cube root
- geometric transformations, including dilations
- basic angle relationships
- volume of spheres and cylinders
- solving linear equations
- determining the number of solutions to a linear equation
- word problems that involve a linear equation
- concept of a function
- concept of a linear function
- rate of change and initial value of a function
- describing functions
- slope and graphing linear equations
- using the Pythagorean Theorem in mathematical and real-world problems
- solving systems of two linear equations
- solving word problems that lead to a system of two linear equations
- associations in scatter plots
- fitting a trend line to a scatter plot; interpreting a trend line for a scatter plot
- associations in two-way tables

Use your judgement in grading. You can give points or partial points for partial answers.

Question #	Max. points	Student score
Exponents and Scientific Notation		
1	8 points	
2	9 points	
3	4 points	
4	2 points	
5	2 points	
	subtotal	/ 25
Irrational Numbers		
6	5 points	
7	5 points	
8	3 points	
9	2 points	
	subtotal	/ 15
Geometry		
10	3 points	
11	2 points	
12	3 points	
13	2 points	
14a	3 points	
14b	3 points	
15	3 points	
16	3 points	
	subtotal	/ 22
Linear Equations		
17	4 points	
18	4 points	
19	6 points	
20	2 points	
21	2 points	
22	3 points	
	subtotal	/21
Functions		
23	2 points	
24a	1 point	
24b	2 points	
24c	2 points	

Question #	Max. points	Student score
Functions		
25a	1 point	
25b	1 point	
25c	1 point	
25d	1 point	
25e	1 point	
26a	2 points	
26b	1 point	
26c	1 point	
26d	1 point	
	subtotal	/17
Graphing Linear Equations		
27a	1 point	
27b	1 point	
27c	2 points	
28	3 points	
29	3 points	
30	3 points	
	subtotal	/13
The Pythagorean Theorem		
31	4 points	
32	3 points	
33	3 points	
	subtotal	/10
Systems of Linear Equations		
34	6 points	
35	3 points	
36	3 points	
37	3 points	
	subtotal	/15
Bivariate Data		
38	3 points	
39	3 points	
40	3 points	
41	5 points	
	subtotal	/14
	TOTAL	/152

Grade 8 End-of-Year Test

Instructions to the student:
Answer each question in the space provided. When applicable, round your answers to a reasonable accuracy according to the context of the problem. A calculator is *not* allowed in the first two sections of the test. A basic calculator (not a graphing calculator) *is* allowed for the rest of the test.

Exponents and Scientific Notation (no calculator allowed)

1. Find the value of the expressions.

a. $-2^4 =$	**b.** $(-2)^4 =$	**c.** $7^{-2} =$	**d.** $6^3 \cdot 6^8 \cdot 6^{-9} =$
e. $31 \cdot 10^{-3} =$	**f.** $10^5 + 10^4 =$	**g.** $\left(\dfrac{-2}{3}\right)^3 =$	**h.** $\dfrac{4^{10}}{4^7} =$

2. Write an equivalent expression using the exponent rules, and without negative exponents.

a. $(-2s)^3 =$	**b.** $(12x)^2 =$	**c.** $(y^3)^5 =$
d. $2x^6 \cdot (-3x^2) =$	**e.** $(y^{-3})^2 =$	**f.** $(4v)^{-2} =$
g. $\left(\dfrac{7x}{3y}\right)^2 =$	**h.** $\left(\dfrac{-x^2}{5x}\right)^3 =$	**i.** $\left(\dfrac{3b}{c^5}\right)^4 =$

3. Write the numbers in scientific notation.

   a. 193 000 000   **b.** 3 080 500 000 000

   c. 0.00046   **d.** 0.0000009

4. The earth's mass is $6.0 \cdot 10^{24}$ kg. Neptune's mass is $1.0 \cdot 10^{26}$ kg.
   What fraction is the earth's mass of Neptune's mass?

5. One gold atom weighs about $3.3 \cdot 10^{-22}$ grams.
   How many gold atoms are in 99 g of gold?

Irrational Numbers (no calculator allowed)

6. Plot the following numbers *approximately* on the number line. Do not use a calculator, but think about between which two whole numbers the root lies, and whether it is close to one of those whole numbers.

    a. $-2\sqrt{2}$ **b.** $\sqrt{82}/3$ **c.** $-\sqrt{10}$ **d.** $\sqrt{35}-1$ **e.** $\sqrt{8}+3$

```
├──┼──┼──┼──┼──┼──┼──┼──┼──┼──┼──┼──┤
-5   -4   -3   -2   -1   0    1    2    3    4    5    6    7
```

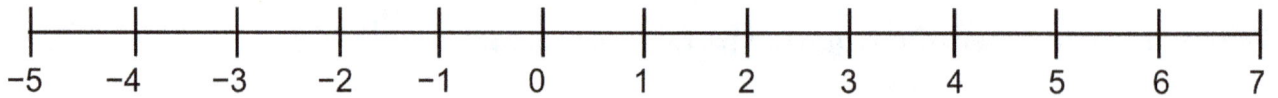

7. Place the numbers in the correct places in the diagram of *real numbers* = the set of both rational and irrational numbers. Note: the set of whole numbers is {0, 1, 2, 3, 4, 5, ...}.

$$9,\ -109,\ \frac{13}{8},\ \sqrt{3},\ \sqrt{49},\ 2\pi,\ 7.89,\ 0.4\overline{1},\ \frac{\sqrt{400}}{8},\ 0,\ -\frac{4}{9},\ \sqrt{2}+1,\ -\frac{35}{7},\ \frac{5}{\sqrt{11}},\ \sqrt{900}$$

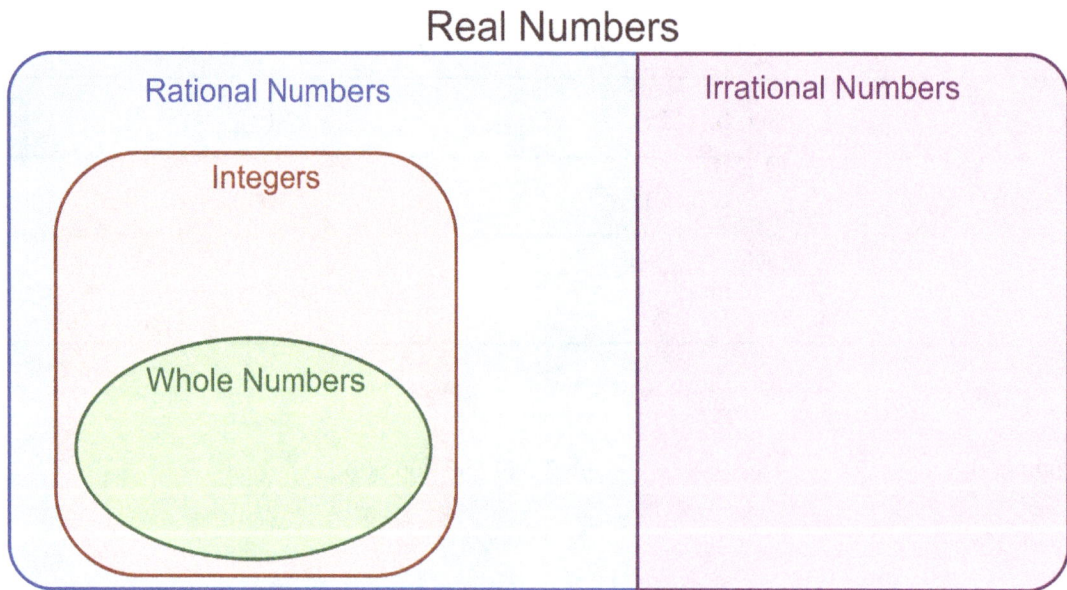

Real Numbers

Rational Numbers	Irrational Numbers
Integers **Whole Numbers**	

8. Solve. If the answer(s) are not rational, give them in root form.

a. $x^2 = 54$	**b.** $3n^2 = 147$	**c.** $z^3 = 64$

9. Write the repeating decimal $0.\overline{71}$ as a fraction.

Geometry

10. Show that the two triangles are similar by describing a sequence of transformations that maps triangle ABC to the smaller triangle.

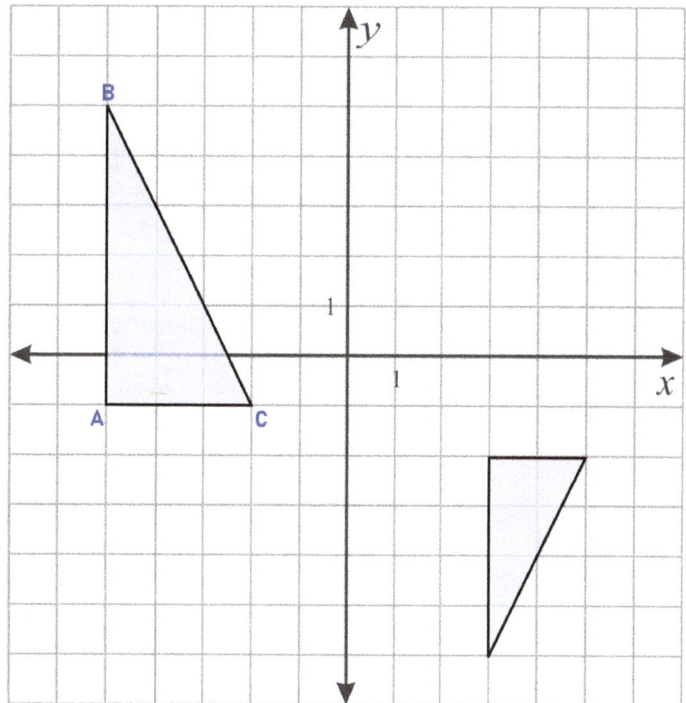

11. Explain a sequence of transformations that can map trapezium ABCD to the smaller trapezium.

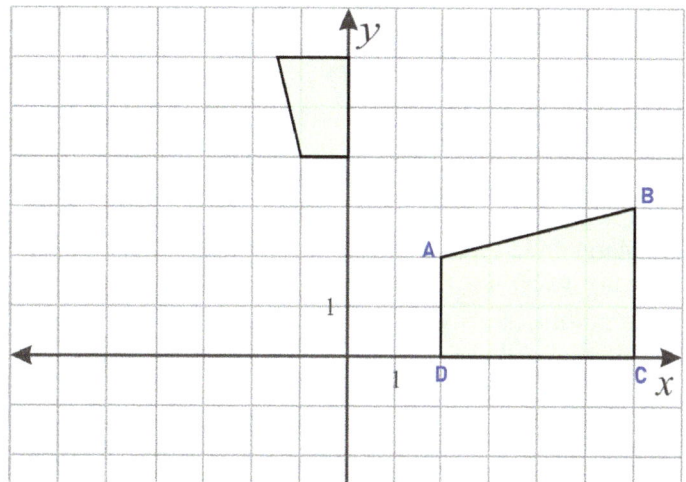

12. Triangle ABC has the vertices A(−2, 5), B(−5, 4) and C(−3, 0). It is transformed in the following ways: First, it is reflected in the y-axis. Then it is translated two units down and one to the left. Lastly, it is rotated 90° clockwise around the origin.

    What are its coordinates after the transformations?

13. Find the value of x, and show your work.
(You can add additional labels to the image, to be
able to reference different parts.)

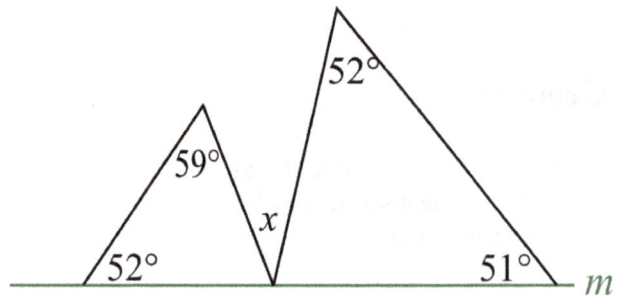

14. **a.** Find the value of x.

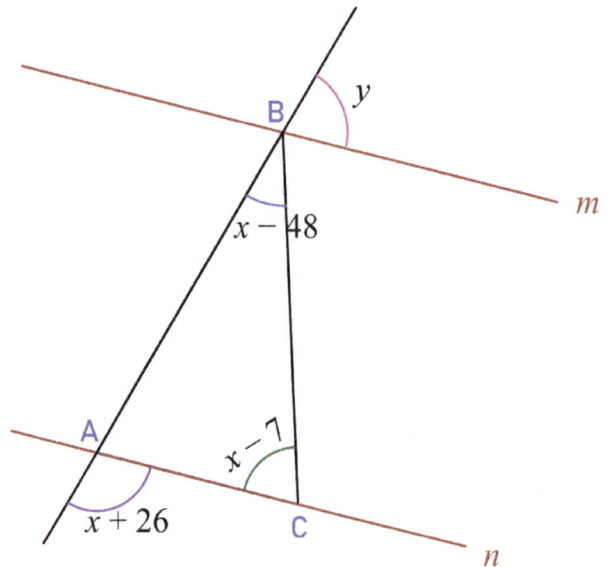

   b. Lines m and n are parallel. Find the value of y.

15. Margaret has a spherical glass vase where she keeps flowers. Its diameter is 15.0 cm. How
much water does she need to fill it 2/3 full? Give your answer both in cubic centimetres and
in millilitres. (Note: 1 L = 1000 cm³.)

16. Lucas is designing a coffee cup in the shape of a circular cylinder. If its interior diameter is 6.2 cm,
what should its (inner) height be, so that its volume would be 340 ml?
Note: 1 ml = 1 cubic centimetre.

Linear Equations

17. Solve.

a. $\quad 10s + 8 \;=\; 7s - 2(s - 5)$	**b.** $\quad 20 - 3(x + 4) \;=\; 14 - 5x$

18. Solve the equations.

a. $\quad \dfrac{2x - 3}{5} - x \;=\; 2$	**b.** $\quad \dfrac{y - 3}{4} \;=\; \dfrac{1 - y}{5}$

19. Solve the equations. Indicate whether each equation has one, none, or an infinite number of solutions.

a. $\quad 6x - 1 \;=\; 6(x - 1)$	**b.** $\quad -5x + 1 \;=\; 6(x - 1) - 5$	**c.** $\quad 6x - 12 \;=\; 6(x - 2)$

53

20. Landon bought 4500 concrete blocks at $1.35 apiece, but he got a discount on a third of them. If his total came to $5775, find how much the discount was.

21. The sum of four consecutive whole numbers is 2342. What are the numbers?

22. The price of an item is reduced by 27%, and then a 6% sales tax is added. You pay $34.82. What was the original price of the item?

Functions

23. **a.** Explain why the graph on the right does *not* depict a function.

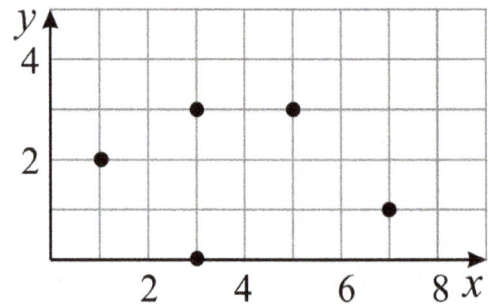

b. Place one of the numbers of 3, 6, or 9 to the empty space in the table so that it is a function.

Input	Output
2	3
5	9
7	5
3	1
	3
9	6

24. Farms A and B both grow strawberries and allow customers to get them for a lower price if they pick them themselves. Farm A charges the customer using the graph below. Farm B uses the table.

Price of Strawberries — Farm A

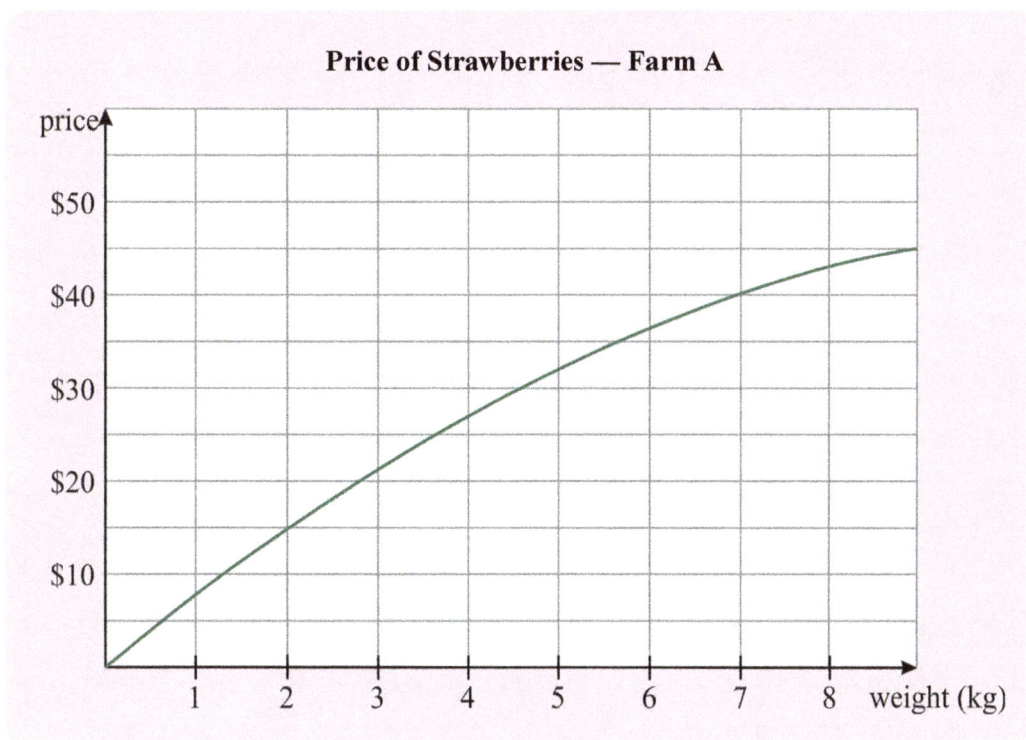

Price of Strawberries — Farm B

Weight (w)	Cost (C)
1 kg	$6.25
2 kg	$12.50
3 kg	$18.75
4 kg	$25.00
5 kg	$31.25
6 kg	$37.50
7 kg	$43.75

a. Which of the two functions is linear? Write an equation for it, using C for cost and w for weight.

b. Find the rate of change for each function from $w = 2$ to $w = 3$.

c. Which farm provides the better deal if you pick 4 kg of strawberries? If you pick 7 kg?

25. The graph below shows the cost for horse riding on a farm as a function of time.

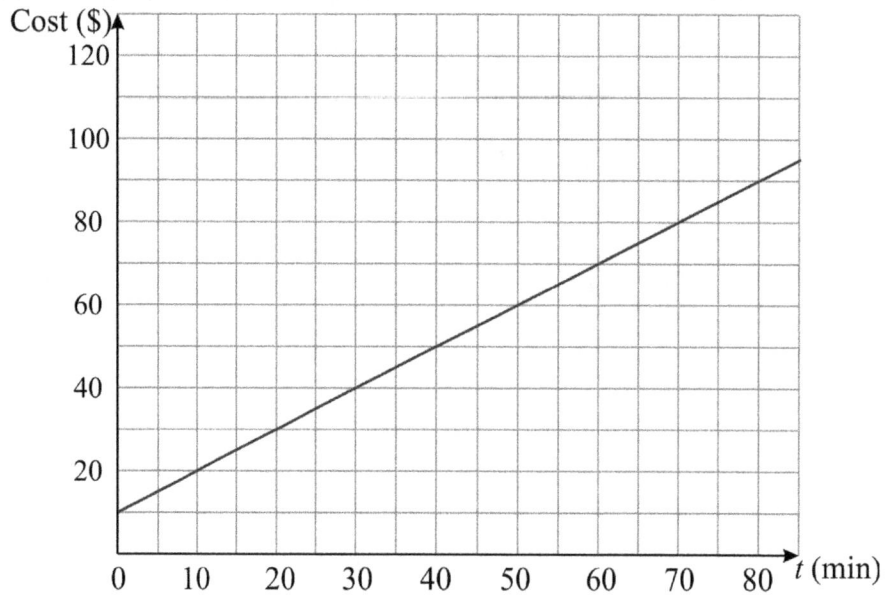

a. What is the initial value of this function?

b. What does it signify in this context?

c. What is its rate of change?

d. What does it signify in this context?

e. Write an equation for the graph.

26. **a.** Describe the function depicted in the graph below by intervals of *x*-values as increasing, decreasing, or constant, and also as linear or nonlinear.

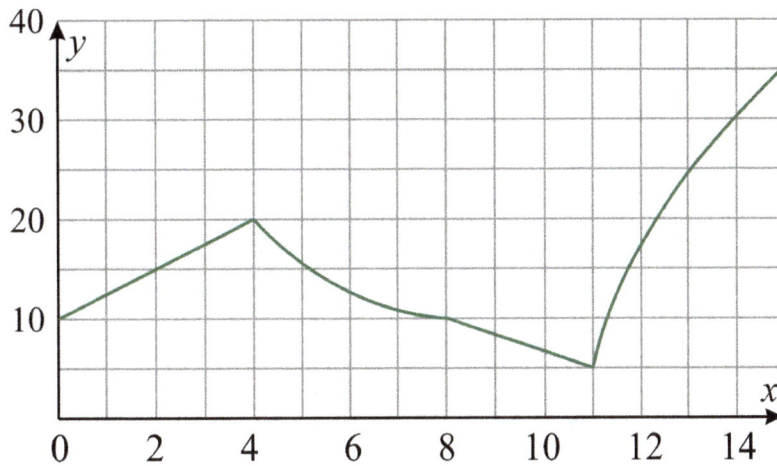

From $x = 0$ to $x = 4$: _____

From $x =$ ____ to $x =$ ____: _____

From $x =$ ____ to $x =$ ____: _____

From $x =$ ____ to $x =$ ____: _____

b. Graph in the same grid a linear function that passes through the points (4, 25) and (8, 20).

c. Find its equation.

d. List the rates of change for both functions from $x = 8$ to $x = 11$.

Graphing Linear Equations

27. Find the equation of each line, in slope-intercept form:

a. has slope −2/3 and passes through (0, 4)

b. is horizontal and passes through (2, −3)

c. has slope 5 and goes through the point (6, 5).

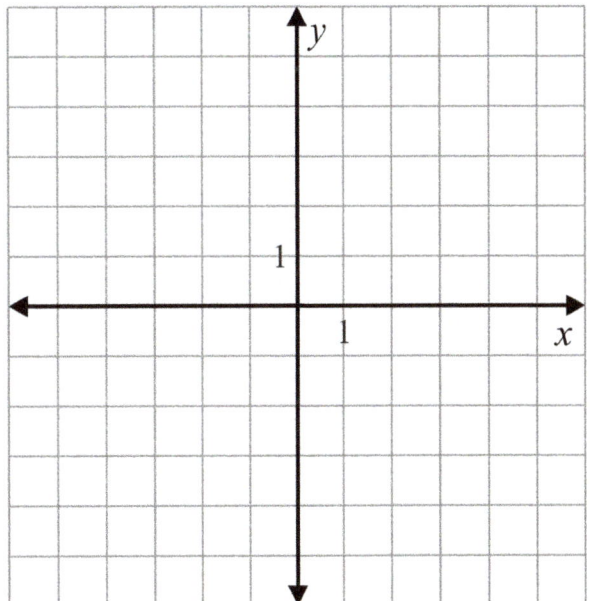

28. Refrigerator companies make estimates of how much energy their fridges consume in typical usage. The table shows how many kilowatt-hours (kWh) of energy Fridge 1 consumed over time, and the graph shows the same for Fridge 2.

Fridge 1

time (mo)	energy (kWh)
3	120
6	240
9	360
12	480

Fridge 2

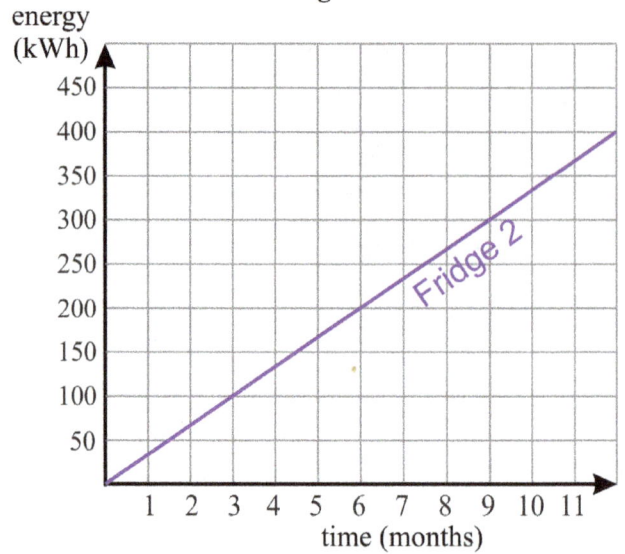

a. Which fridge consumes more electricity in *three* months?

How much more?

b. Write an equation for each fridge's energy consumption, relating the energy (E, in kWh) and the time (*t*, in months).

c. Plot the equation for Fridge 1 in the grid.

29. Graph the lines.

a. $y = (-1/2)x - 3$

b. $2x - 3y = 6$

c. $x = -4$

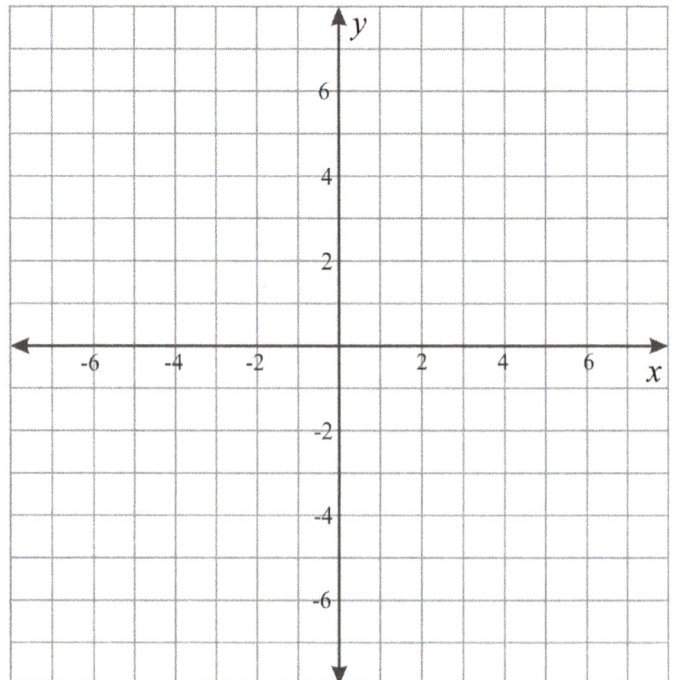

30. Find a so that the point $(a, 2)$ will fall on the same line as the points $(3, 14)$ and $(-7, -6)$.

The Pythagorean Theorem

31. Solve for the unknown side.

a.	b.
r 17.5 26.6	$\sqrt{70}$ x x

32. How long is the rafter (the roof piece) in this chicken coop design?

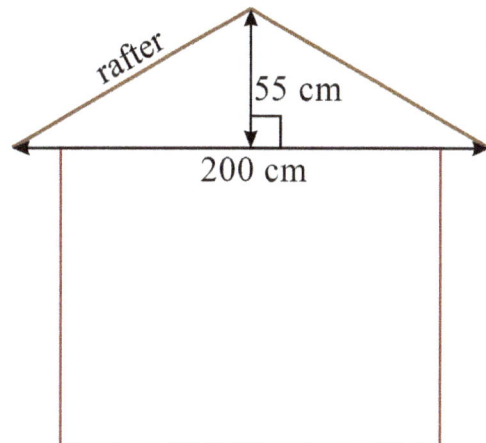

rafter   55 cm   200 cm

59

33. **a.** Find the height of this square pyramid.

b. Find its volume.

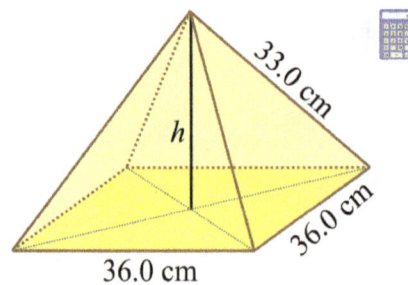

Systems of Linear Equations

34. Solve each system of equations. Give the solutions as fractions.

a. $\begin{cases} 2x - 3y = 8 \\ 3x + 4y = -5 \end{cases}$

b. $\begin{cases} -x = 4(y + 5) \\ 2x = -12y - 10 \end{cases}$

35. Tell how many solutions each system of equations has by inspecting the equations. You do not have to find the solution(s).

a. $\begin{cases} 4x - 2y = -8 \\ 2y - 4x = 1 \end{cases}$

b. $\begin{cases} 4x - 2y = -1 \\ -4y + 2x = 2 \end{cases}$

c. $\begin{cases} 4x - 2y = -1 \\ 2y - 4x = 1 \end{cases}$

36. A restaurant has two kinds of tables: ones that seat 4, and ones that seat 6. They have a total of 106 tables (some are in storage), and can seat 500 people.

Write a system of two equations to model this situation, and solve it.

37. Greta said to Susan, "In ten years, my age will be 3/4 of your age." If the sum of their ages is 127, find their ages now.

Bivariate Data

38. Explain whether each scatterplot shows an association between the variables. If yes, classify the association as linear or nonlinear, increasing or decreasing.

a.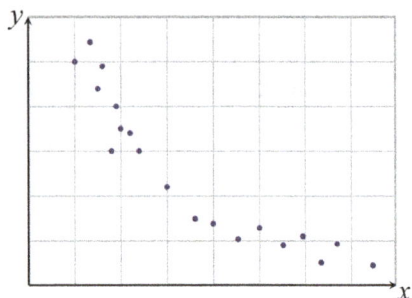

b.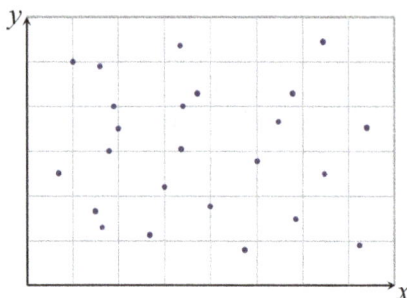

c.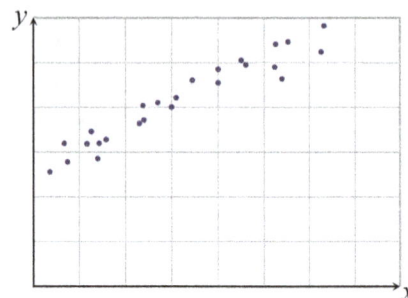

39. Is there an association between the two variables? Also, explain how you know that.

Age	Exercises	Does not exercise	Total
15-24	51	47	98
25-34	52	52	104
35-44	52	47	99
45-54	28	24	52
55-64	22	27	49
TOTAL	205	197	402

40. Which of the following numbers of 3, 6, or 24, would you put in the empty box so that...

a. there is no association between the variables?

b. children ages 9-11 are far more likely to walk to school than children ages 6-8?

c. children ages 6-8 are far more likely to walk to school than children ages 9-11?

	Walks to school?	
	yes	no
Ages 9-11	20	10
Ages 6-8	12	

41. **a.** Draw a line to fit the trend in the scatter plot below.

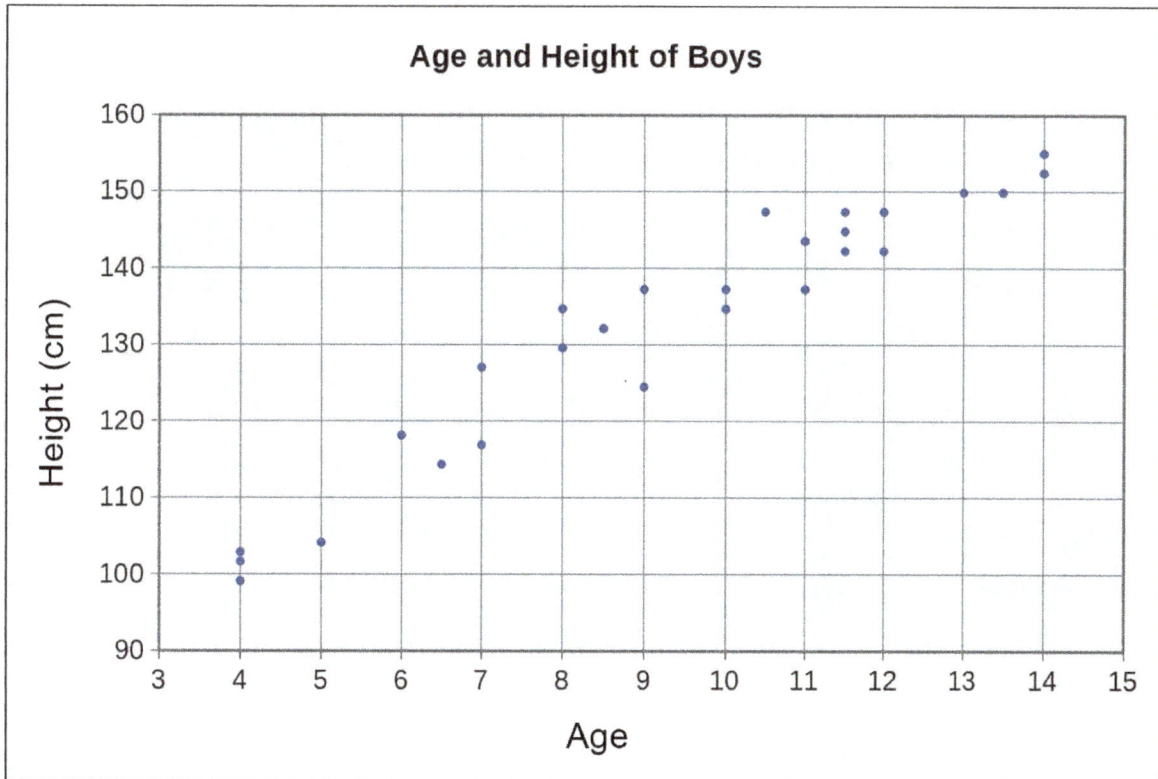

Age and Height of Boys

b. Find the (approximate) equation of *your* line.

c. A spreadsheet program calculates the equation for a trendline to be approximately $y = 5.3x + 82.9$. What does the slope of that equation signify in this context?

d. What does the *y*-intercept of that equation signify in this context?

e. What age does the equation from (c) predict for a boy that is 128 cm tall?

Using the Cumulative Revisions

The cumulative revisions practise topics in various chapters of the Math Mammoth complete curriculum, up to the chapter named in the revision. For example, a cumulative revision for chapters 1-6 may include problems matching chapters 1, 2, 3, 4, 5, and 6. The cumulative revision lesson for chapters 1-6 can be used any time after the student has studied the curriculum through chapter 6.

These lessons provide additional practice and revision. The teacher should decide when and if they are used. The student doesn't have to complete all the cumulative revisions. I recommend using at least three of these revisions during the school year. The teacher can also use the revisions as diagnostic tests to find out what topics the student has trouble with.

Math Mammoth complete curriculum also includes an easy worksheet maker, which is the perfect tool to make more problems for children who need more practice. The worksheet maker covers most topics in the curriculum, excluding word problems. Most people find it to be a very helpful addition to the curriculum.

You can access the worksheet maker online at

https://www.mathmammoth.com/private/Make_extra_worksheets_grade8.htm

Cumulative Revision, Grade 8, Chapters 1-2

1. The points $(-5, -1)$, $(-3, 3)$, and $(-1, 0)$ are the vertices of a triangle. It is rotated 90° clockwise around the origin, and then translated 2 units to the left and 3 units up. What are the coordinates of its vertices now?

2. A rectangle is first moved 3 units down and 2 to the left, then reflected in the x-axis. Its vertices are now at $(12, -6)$, $(12, -10)$, $(9, -6)$, and $(9, -10)$.

   What were its coordinates before these transformations?

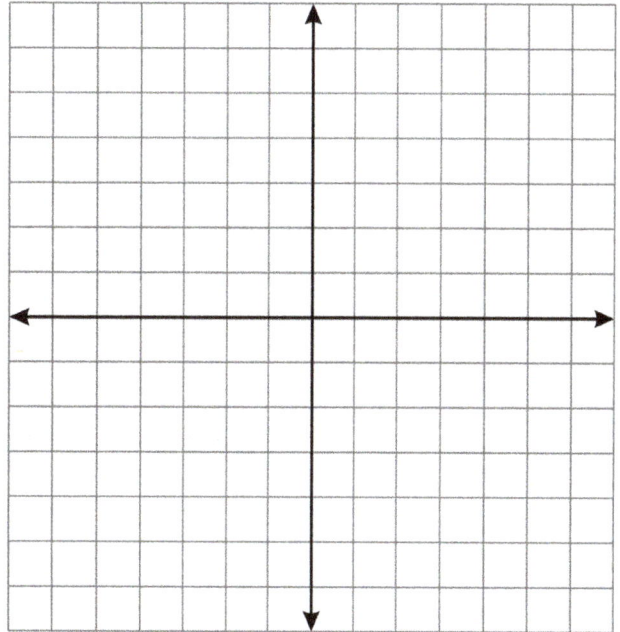

3. Draw dilations of triangle ABC according to the instructions.

   a. Draw a dilation of triangle ABC using origin as centre, and the scale factor of 1/2.

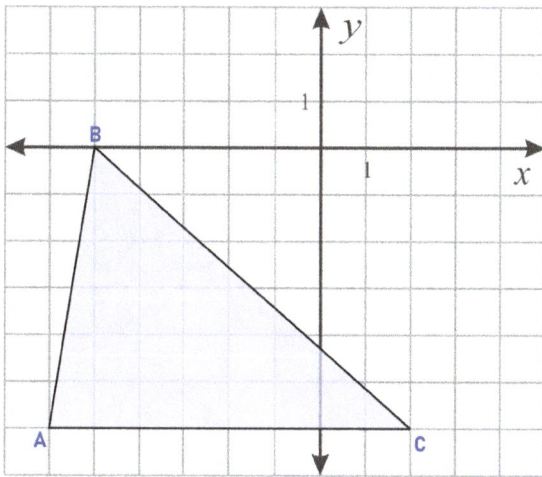

   b. Draw a dilation of triangle ABC from point B, again using the scale factor of 1/2.

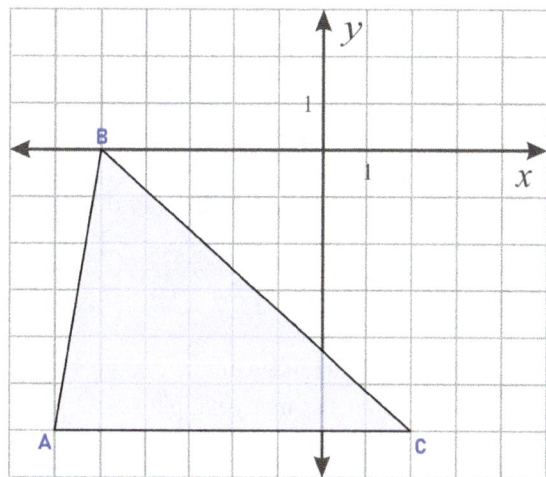

4. Find the value of the expressions. Do not use a calculator.

a. $-2^4 =$	**b.** $(-3 \cdot 2)^{-2} =$	**c.** $3 \cdot 10^{-3} =$	**d.** $5^{-2} \cdot 5^4 \cdot 5^{-3} =$
e. $(3 \cdot 10)^{-3} =$	**f.** $3^2 + 3^3 + 3^4 =$	**g.** $\left(\dfrac{-5}{6}\right)^2 =$	**h.** $\dfrac{11^6}{11^8} =$

5. Simplify, and write an equivalent expression that does not have any negative exponents.

a. $(5s^2)^{-1} =$	**d.** $(xy^{-3})^5 =$	**g.** $(2a)^3 \cdot a^{-2} =$
b. $(4t^{-3})^{-2} =$	**e.** $(3xy^3)^{-2} =$	**h.** $(-10y)^{-3} \cdot (4y)^2 =$
c. $(xy^3)^{-2} =$	**f.** $(-2w^{-2})^4 =$	**i.** $(-3x)^{-1} \cdot (3x)^3 =$

6. Find the value of each unknown.

a. $(5^x)^2 = 25^4$	**b.** $16^3 = (4^y)^2$	**c.** $(16 \cdot 3^x)^{-2} = \dfrac{1}{6^8}$

7. **a.** Match the expressions and numbers with the same value.

   b. Circle the ones that are written in scientific notation correctly.

$0.07 \cdot 10^9$	$7 \cdot 10^4$	$7 \cdot 10^9$
$70\,000\,000$	$70 \cdot 10^8$	$0.7 \cdot 10^5$
$0.7 \cdot 10^8$	$7\,000\,000\,000$	$70\,000$

8. Rewrite in scientific notation correctly.

   a. $0.93 \cdot 10^7$ **b.** $23 \cdot 10^6$ **c.** $14.5 \cdot 10^8$

9. A student multiplied two large numbers with a calculator and got this: $\boxed{2.1\text{E}25}$

   a. What does the answer mean?

   b. What two numbers could she have multiplied?

10. The population of India is estimated to be $1.5 \cdot 10^9$ in 2030. That is $2 \cdot 10^4$ times as many people as in Lydia's hometown.

    What is the population of Lydia's hometown?

11. Calculate with a calculator. Round your answer to the correct amount of significant digits.

a. 8.04 m $\cdot$ 2.5 m	**b.** $250 \cdot 0.493$ kg
c. $(4.71 \cdot 10^4$ dollars$) \div 1944$ hr	**d.** 49 L $\div 757$ km

Cumulative Revision, Grade 8, Chapters 1-3

1. Show that the two pentagons are similar by describing a sequence of transformations that maps the larger pentagon to the smaller one.

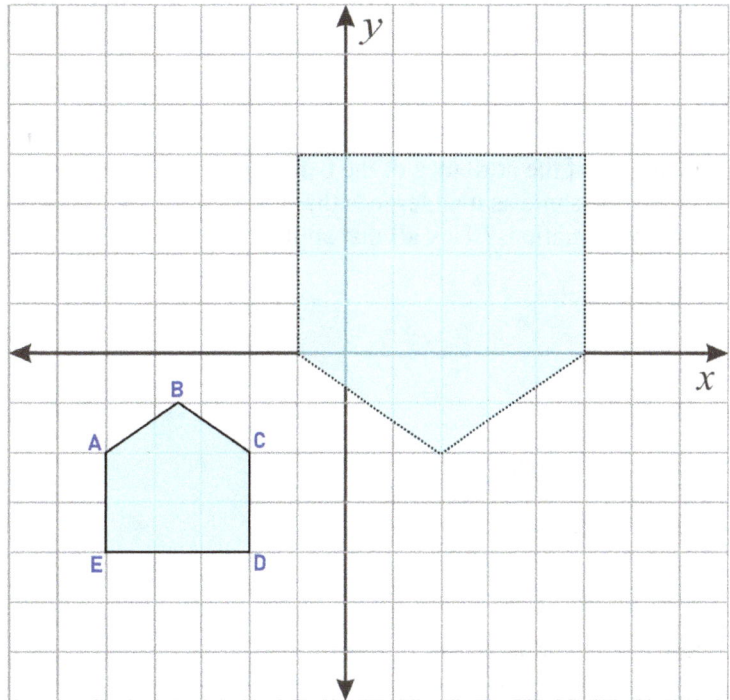

2. Draw a dilation of this triangle from point C and with a scale factor of 1/2.

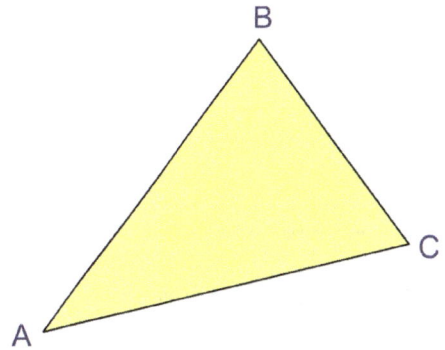

3. Lines m and n are parallel. Find the measure of the angle marked with "?". Explain your reasoning.

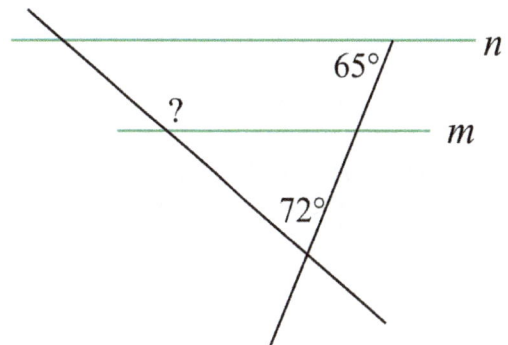

4. Parallelogram ABCD undergoes two congruent transformations, as shown in the image on the right.

   a. Name the two transformations.

   b. Which of the attributes of the parallelogram stay the same as it undergoes the transformations? Tick all that apply.

   (i) Perimeter

   (ii) Area

   (iii) Position

   (iv) Angle ABC

   (v) Angle sum

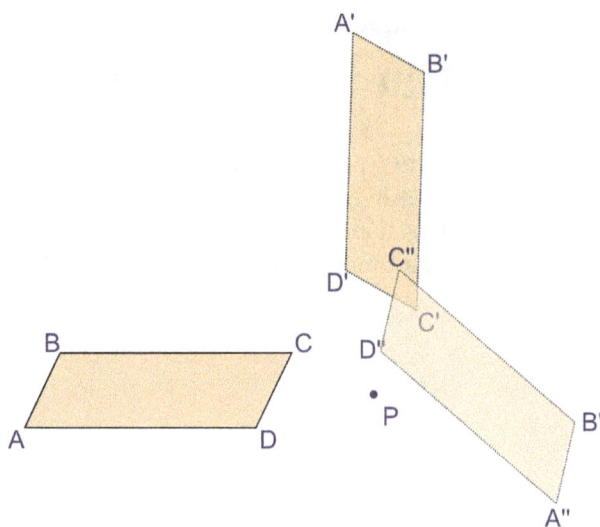

5. Solve. Give your answer in scientific notation.

a. $2 \cdot 10^8 + 8 \cdot 10^7$	**b.** $1.3 \cdot 10^6 + 2 \cdot 10^7$	**c.** $3 \cdot 10^5 - 9 \cdot 10^4$	**d.** $2.6 \cdot 10^8 - 5 \cdot 10^7$

6. One of these simplification processes has an error. Which one? Correct it.

a. $\left(\dfrac{4a}{-b^2} \right)^2 = \dfrac{(-4a)^2}{(-b^2)^2} = \dfrac{16a^2}{b^4}$.	**b.** $(2 \cdot 10)^{-3} = \dfrac{2}{10^3} = \dfrac{2}{1000}$

7. A dump truck just dumped 8.41 cubic metres of sand on the ground, and it is in a form of a circular cone now. The diameter of the bottom circle is 3.66 m. How tall is the cone of sand?

8. Calculate, and give your answer to a reasonable accuracy.

    a. 12.5 km + 2.2 km + 0.054 km = _____ km   **b.** 4.6 km + 409 km + 34 km = _____ km

9. The distance from a campsite to a river is 4.2 cm on a map with a scale of 1:4000.
   How long is this distance in reality, in metres? (Take note of the significant digits in your answer.)

10. How many times does a person's heart beat, in an average lifetime of 70 years? Assume that the
    average heart rate is 60 beats per minute. Give your answer in scientific notation, and to two
    significant digits.

11. The Proxima Centauri star is $4.0208 \cdot 10^{13}$ km away from Earth. The speed of light is $3.00 \cdot 10^{5}$ km/s.
    How long does the light from this star take to reach Earth? Give your answer in years.

Cumulative Revision, Grade 8, Chapters 1-4

1. Figure DEFG underwent a dilation, then a reflection. Study the coordinates to find out the details about each transformation, then fill in the missing coordinates.

Original figure	Dilation	Reflection
D(−5, −4)	D'(___ , ___)	D"(−2.5, 2)
E(−6, −2)	E'(−3, −1)	E"(___ , ___)
F(−1, −2)	F'(___ , ___)	F"(−0.5, 1)
G(−2, −4)	G'(−1, −2)	G"(−1, 2)

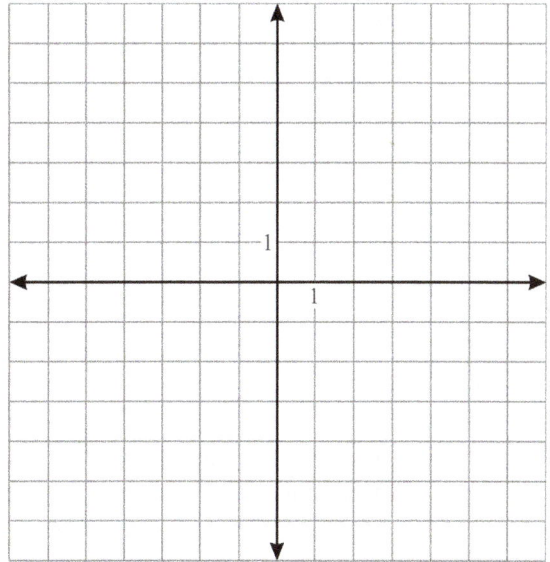

2. A kite was first rotated 180° around the origin, then translated three units to the right, and lastly reflected in the x-axis. Now its vertices are at points (6, 4), (6, 6), (4, 6), and (2, 2).

   a. What were the coordinates of its vertices before these transformations?

   b. What single transformation would have produced the same result?

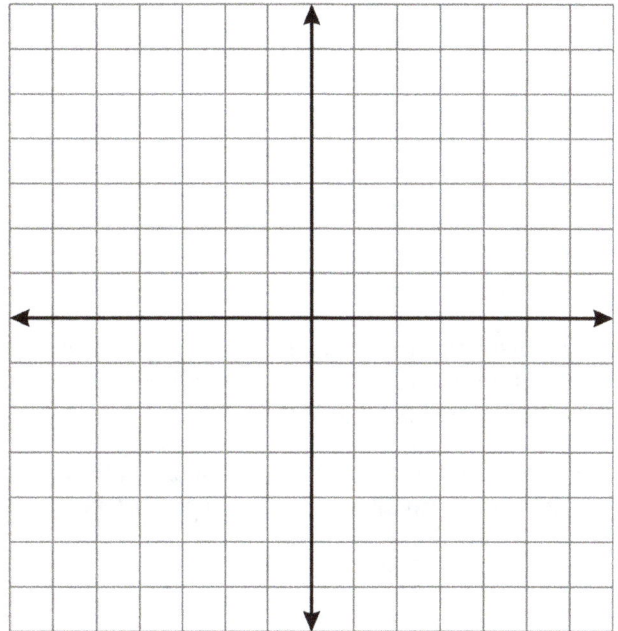

3. An item had three price increases: for 4%, for 7.5%, and for 5%. Now it costs $136.45. What was its price before these increases?

4. Lines *m* and *n* intersect at point P. Are the triangles ABP and PQR similar? Explain your reasoning.

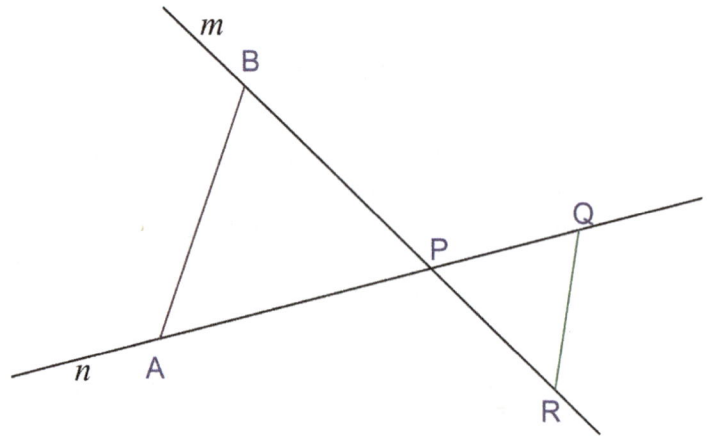

5. Line *m* is parallel to the line segment $\overline{AB}$. Find the measures of angles *x* and *y*.

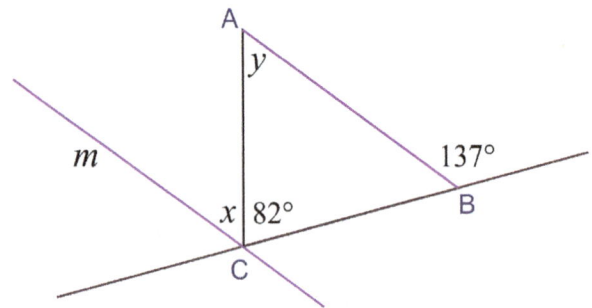

6. You and your friend are in a hardware store, and you are considering buying a low-priced 15-litre bucket, in the shape of a circular cylinder. Your friend says, "I doubt 15 L of water can fit in that thing; they're cheating you."

You have a tape measure so you quickly measure the diameter of the bottom (it is 28.0 cm) and the height (it is also 28.0 cm). Should you believe your friend? Explain your reasoning.

Note: $1 \text{ cm}^3 = 1 \text{ ml}$

7. Convert. Remember to round your answer to the same amount of significant digits as the measurement.

a. 230 cm = _____ in

b. 54 L = _____ gal

c. 24.5 ft = _____ m

d. 437 in = _____ m

1 inch = 2.54 cm

1 ft = 0.3048 m

1 gal = 3.785 L

8. One molecule of water consists of two hydrogen atoms and one oxygen atom (H_2O). The mass of one water molecule is about $2.99 \cdot 10^{-23}$ grams. How many water molecules are in 100 grams of water?

9. What errors are made in these solutions? Correct them, and continue the solutions.

a. $5 + x + \dfrac{x-3}{2} = 6 \qquad \Big| \cdot 2$

$10 + x - 3 = 12$

b. $x - 2 = 2x - \dfrac{x-1}{5} \qquad \Big| \cdot 5$

$5x - 10 = 10x - x - 1$

10. The formula $m = \dfrac{a_1 + a_2 + a_3 + a_4}{4}$ gives you the mean (average) of four numbers a_1, a_2, a_3, and a_4.

   a. Solve this formula for a_2.

   b. Robert has gotten 67, 85, and 76 points on three maths tests, and he has one more to go. He wants his average for the three tests to be 80 (at least). What should he get on his fourth test in order to achieve that?

11. The sum of three consecutive even whole numbers is 13 788. What are the numbers?

Cumulative Revision, Grade 8, Chapters 1-5

1. For what value of a would the equation $4(x + 7) = 10 - a(2x - 1)$ have *no* solutions?

2. The area of a kite is $A = \dfrac{pq}{2}$, where p and q are the diagonals of the kite.

   a. Solve this formula for p.

   b. If the area of a kite is 0.6 m^2, and the shorter
   diagonal is 90 cm, how long is the other diagonal?

3. **a.** Ann made a circular cake (in the form of a circular cylinder). Its bottom diameter is 23 cm and
   its height is 10 cm. If it is divided into 12 equal slices, what is the volume of one slice?

   b. If a square cake is baked in a 20 cm by 20 cm pan, what should its height
   be so that it would have the same volume as the circular cake in the previous question?

4. Solve.

a. $2x = \dfrac{x - 5}{6} - 2$	**b.** $\dfrac{x - 1}{4} + \dfrac{2x + 7}{3} = 0$

5. An aeroplane travels at a constant speed of 1000 km/h from New York to Los Angeles, a distance of 3940 km.

   a. Consider the distance (d) the aeroplane has travelled as a function of time (t).
   Write an equation relating the two variables.

   b. What is the rate of change of this function?

   c. Is this function linear? How do you know?

   d. Plot your equation. Notice that you need to decide a scale for the d-axis.

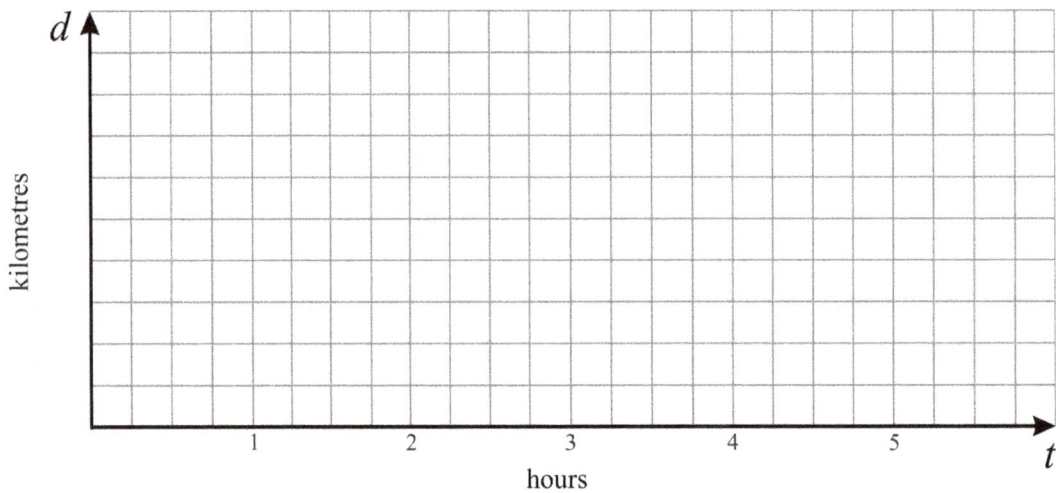

   e. How far will the aeroplane travel in 1 hour 40 minutes?

6. The table shows the distance as a function of time for another aeroplane.

   a. Is this function linear?

   How do you know?

   b. Compare this to the function in question #5.
   Which function has a greater rate of change from 3 to 4 hours?

Time (h)	Distance (km)
0	0
0.5	450
1	900
1.5	1425
2	1800
2.5	2250
3	2625
3.5	3150
4	3600

7. Simplify, writing an equivalent expression that does not have any negative exponents.

a. $(2x^3)^{-1} =$	c. $(ab^3)^{-2} =$	e. $(-2x)^5 \cdot (5x)^{-1} =$
b. $(8a^{-5})^2 =$	d. $(-3s^{-2}t)^3 =$	f. $z^4 \cdot (-3z^3)^{-2} =$

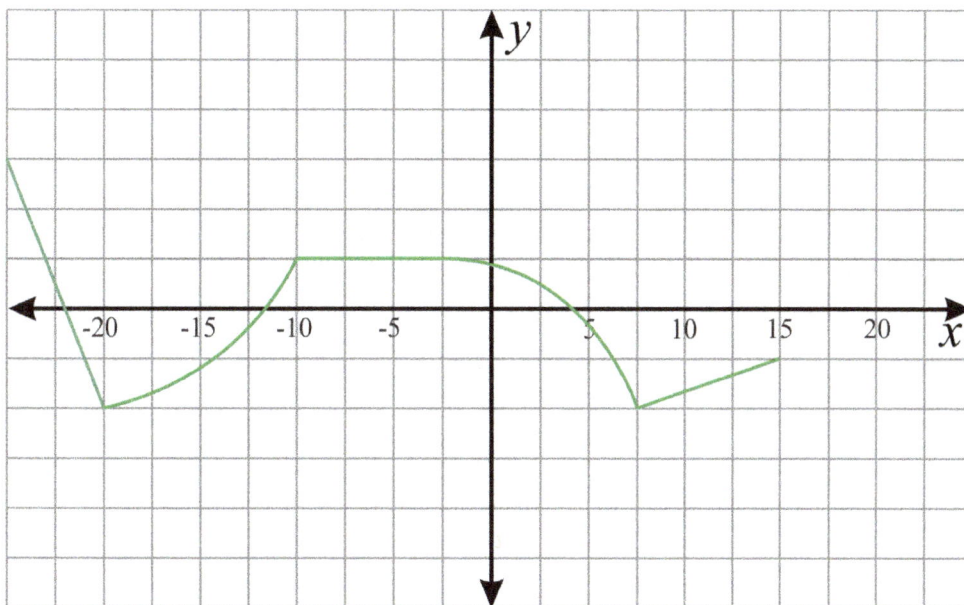

8. Above, you see the graph of a certain function. Describe where this function is

    a. linear and decreasing

    b. constant

    c. nonlinear and decreasing

9. Continue drawing the graph of the function in #8 from $x = 15$ till $x = 25$ so that it is nonlinear and increasing.

Cumulative Revision, Grade 8, Chapters 1-6

1. Solve.

a. $2 - \dfrac{x+1}{5} = x$	**b.** $4y + \dfrac{1-2y}{10} = 3y$
c. $\dfrac{x-10}{3} = \dfrac{3x+4}{5}$	**d.** $-x + \dfrac{7x-3}{4} = \dfrac{x}{2} - 2$

2. Rewrite the numbers in scientific notation correctly.

   a. $72 \cdot 10^{-6}$ **b.** $233 \cdot 10^{6}$

   c. $0.04 \cdot 10^{-2}$ **d.** $0.309 \cdot 10^{9}$

3. Compare the numbers, writing < , >, or = in the box.

   a. $1.9 \cdot 10^{8}$ ☐ $9.1 \cdot 10^{7}$ **b.** $9 \cdot 10^{-11}$ ☐ $9 \cdot 10^{-12}$ **c.** $5.2 \cdot 10^{-5}$ ☐ 0.000053

   d. $9 \cdot 10^{-4}$ ☐ 0.0009 **e.** $7 \cdot 10^{4}$ ☐ $100\,000$ **f.** 0.00000053 ☐ $1 \cdot 10^{-6}$

4. Tell, without solving the equations, whether each equation has one unique solution, no solutions, or an infinite number of solutions.

    a. $-3x + 6 = -3x - 9$

    b. $-3x + 2 = -1$

    c. $-3x + 4 = 4$

    d. $-3x = 5 - 3x - 5$

5. Choose the relationships that are functions.

(1)

Input	Height (cm)	131	130	125	135	131
Output	Name	Henry	Harry	Helen	Hal	Holly

(2) Let S be a rule that takes any positive number x as input, and gives $1/x$ as output.

(3) Input is a person's social security number, output is their name.

(4) Input is a date (such as July 8), output is the person whose birthday it is.

6. Three functions are represented below.

    a. Which one has the largest initial value?

    b. Which one(s) are linear functions?

    c. Which one has the smallest rate of change in the x-interval [6, 10]?

    d. Describe each function in the x-interval [4, 6] as increasing, decreasing, or constant.

Function 1:

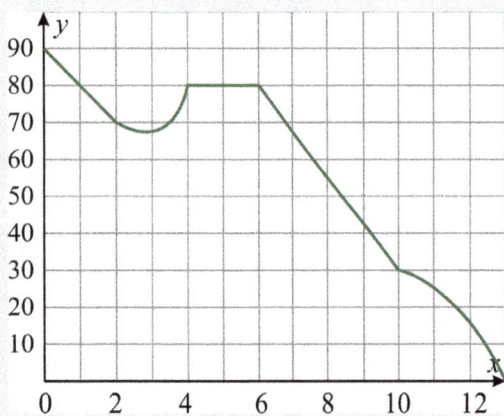

Function 2:

$y = 6.5x + 10$

Function 3:

x	y
0	80
2	65
4	50
6	35
8	20
10	5
12	−10

7. Grace deposits $5000 in a savings account with a 6% interest. The interest will be added to the principal every year (compound interest). She is told that the final amount (A) on her account after t years is given by the formula $A = 5000(1.06)^t$. Is this function a linear function? Explain.

8. Construct two functions below by filling in the rest of the output values. In (a), the function should be a *linear* function. In (b), it should be *nonlinear*.

a.

Input (x)	−5	−4	−3	−2	−1	0	1	2	3	4	5
Output (y)					0	3	6				

b.

Input (x)	−5	−4	−3	−2	−1	0	1	2	3	4	5
Output (y)	63	56	49	42	35						

9. Find the volume of a square pyramid with a 30-cm by 30-cm bottom square, and a height of 24 cm.

10. Karen wants to do some remodelling in her kitchen. She is comparing the pricing from two companies. Company 1 charges $42 per work hour plus a fixed fee of $1800 for the materials. Company 2 will charge $36 per work hour and $2400 for the materials.

   a. Write an equation for the cost (C) of using Company 1 as a function of work hours (h).
   Do the same for Company 2.

   Company 1: _____   Company 2: _____

   b. If it is estimated that the job will take 15 hours, which company is a better deal?

   c. For what number of work hours does Company 1 cost the same as Company 2?

Cumulative Revision, Grade 8, Chapters 1-7

1. Write in order from the smallest to greatest value: $2 \cdot 10^{-5}$ $\quad -2 \cdot 10^{5}$ $\quad 0.0005$ $\quad -2 \cdot 10^{-5}$ $\quad 2 \cdot 10^{5}$

2. Are the triangles similar? Explain how you know.

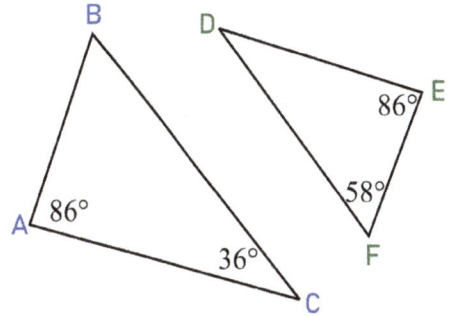

3. The table shows the cost of renting a snowboard as a function of time.

Days	Cost ($)
0	10
1	65
2	120
3	175
4	230
5	285

   a. Write an equation to model this (linear) relationship.

   b. What is the rate of change of this function?

   What does that signify in this context?

   c. What is the initial value of this function?

   What does that signify in this context?

4. Sketch a graph to depict the distance of the boat from the shore as a function of time.

   The boat starts out far away from the shore. It approaches the shore at a steady speed for two days.
   Then it advances towards the shore at a greater speed for one day, and reaches it.
   It stays docked for two days. Then it goes out to the ocean again, its distance from the shore growing in a linear manner, for three days. After that, its distance from the shore continues to increase, but gradually slows down. For day 10, it stays in its place.

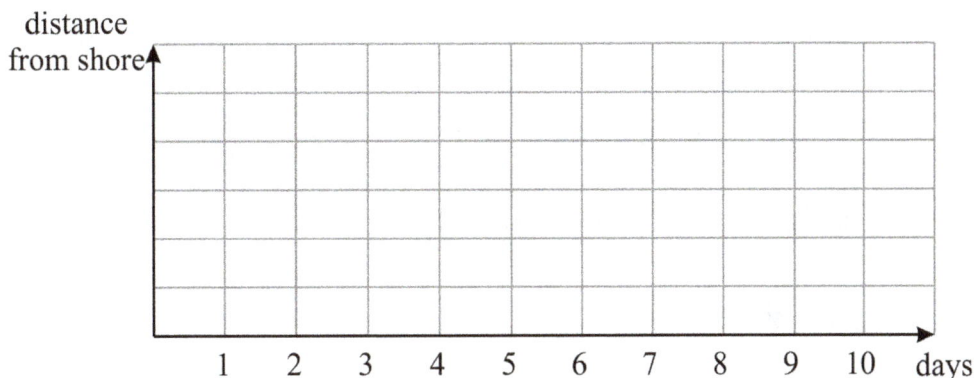

5. The perimeter of an isosceles triangle is 65 units. If its base side was 7 units longer, the triangle would be equilateral. Find the sides of the isosceles triangle.

6. ABCD is a parallelogram. Find the value of x.

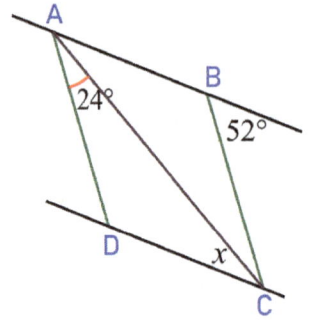

7. **a.** Draw a line with a slope of 3/2 that goes through the point (0, −2). Then write its equation.

   b. Draw a line with a slope of −3 that goes through the point (−4, 5). Then write its equation.

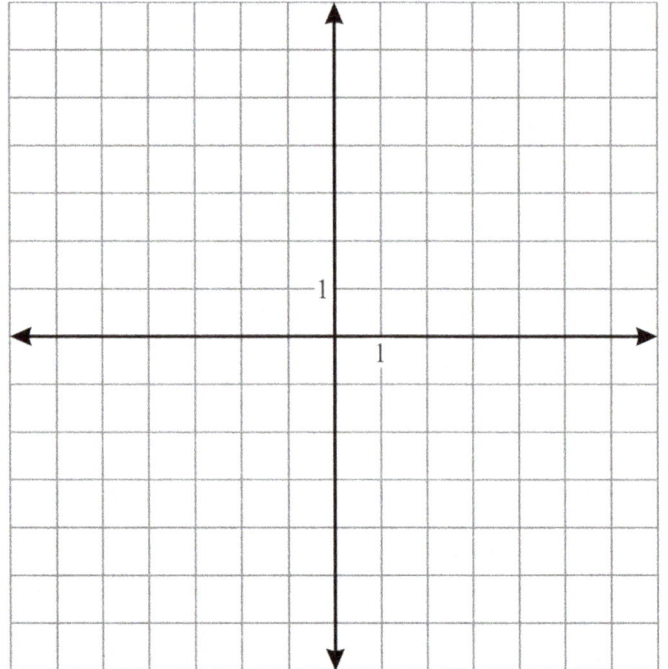

8. Match the descriptions and the equations.

a. Passes through (4, 1) and has slope 1/4

(i) $y = (1/4)x - 2$

b. Passes through (−1/4, 1/4) and has slope 0

(ii) $x = 1/4$

c. Passes through (1/4, 1/4) and has no slope

(iii) $y = (1/4)x$

d. Passes through (4, −1) and (−8, −4)

(iv) $y = 1/4$

9. For each line given in standard form, find its x and y-intercepts. Then graph the line.

a. $4x - 2y = 8$

b. $9x + 3y = -18$

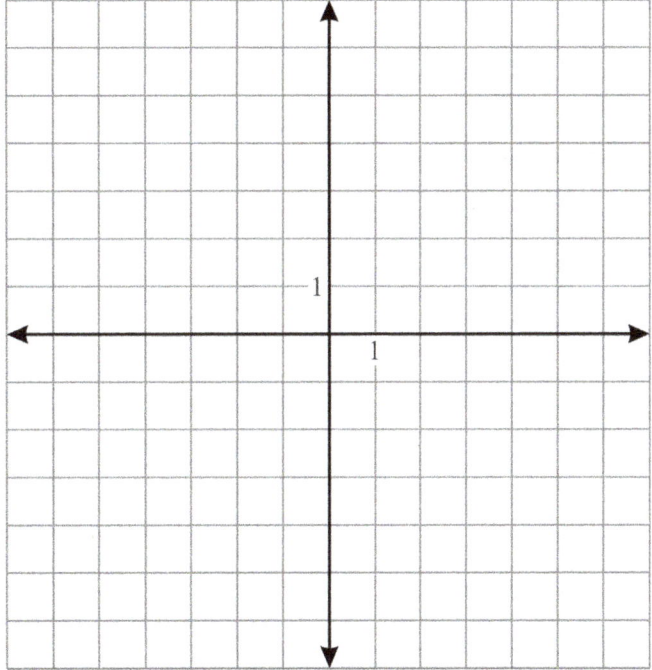

10. An item had three price increases: for 3%, for 4%, and for 9%. Now it costs $268.55. What was its price before these increases?

11. Often, drinking glasses are in the form of a cut cone. Calculate the volume of the glass on the right, using this technique:

(1) Calculate the volume of the entire cone shape that is 25 cm tall.
(2) Calculate the volume of the cut part, that is also a cone (at the bottom).
(3) Subtract the above two results.

(The image is not to scale.)

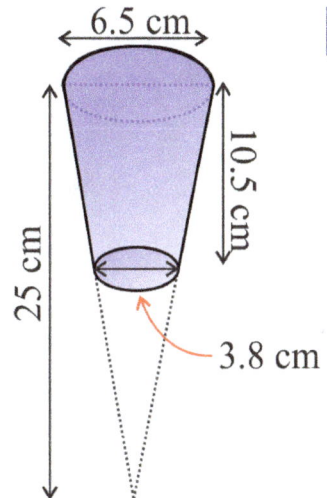

Cumulative Revision, Grade 8, Chapters 1-8

A basic calculator (not a graphing calculator) is allowed for all the exercises.

1. The points A(5, −1), B(3, −4), and C(1, −3) are the vertices of a triangle. It is first reflected in the *x*-axis. Then it is dilated from the origin with a scale factor of 2. Lastly, it is rotated 90° clockwise around the origin. What are the coordinates of its vertices now?

2. **a.** How many solutions does this equation have?

   $$10s - 6 = 2s$$

   b. Modify only one term in the equation so that the resulting equation will have no solutions.

3. Solve the systems of equations.

   a. $\begin{cases} 3x - y = 14 \\ 6x + 3y = -12 \end{cases}$

   b. $\begin{cases} 3x - 8(y - 2) = 0 \\ 2x = 8y + 1 \end{cases}$

4. Some of the statements below are incorrect. Correct them.

   a. $\sqrt{121}$ is irrational because it is a square root.

   b. 0.831831831 is rational because its decimal expansion repeats.

   c. $\sqrt{13}$ is irrational because it is a square root of a non-perfect square.

   d. $\dfrac{\pi}{6}$ is irrational because π is irrational, and an irrational number divided by a rational one is irrational.

   e. $\dfrac{\sqrt{63}}{4}$ is rational because it is a fraction.

5. How much milk that is 3% butterfat and cream that is 15% butterfat should you mix in order to get one litre of "milky cream" that is 12% butterfat?

   Filling in the chart can help.

	volume	butterfat percentage	butterfat amount
3% milk			
15% cream			
Mixture			

6. Find the error in the solution of this system of equations. Then correct the error and solve the system.

(1) $\quad \begin{cases} 3x - 7y = 4 \\ -5x + y = -4 \end{cases} \Big| \cdot 7$
(2)

$\downarrow$

$\begin{cases} 3x - 7y = 4 \\ -35x + 7y = -28 \end{cases}$

$\underline{}$

$\qquad -32y = -24$

$\qquad\quad y = 3/4$

$\downarrow$

(1) $\quad 3x - 7(\mathbf{3/4}) = 4$

$\qquad 3x - 21/4 = 4$

$\qquad\qquad 3x = 4 + 21/4 = 37/4$

$\qquad\qquad 3x = 37/4$

$\qquad\qquad\ x = 37/12$

However, (37/12, 3/4) does *not* fulfill the 2nd equation.

90

7. Solve for the unknown side. In (a), give your answer to one decimal. In (b), give the exact answer.

a.

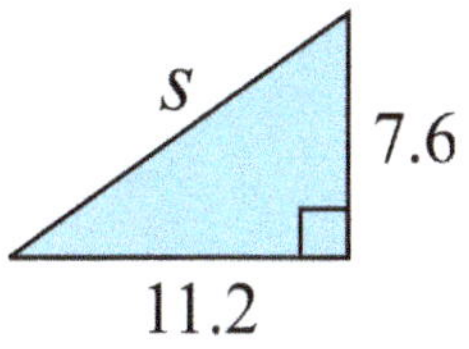

s

7.6

11.2

b.

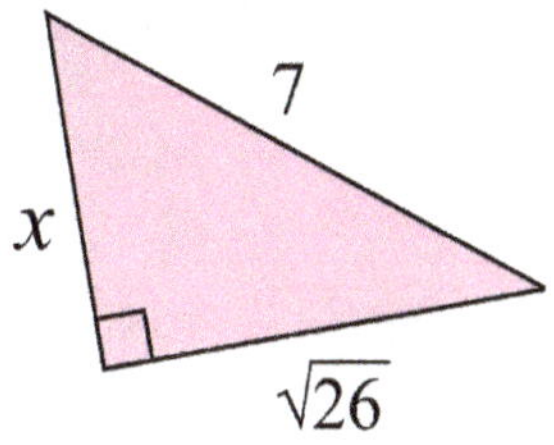

7

x

$\sqrt{26}$

8. Calculate the area of an equilateral triangle with 40.8-cm sides.
   Don't forget to draw a sketch.

Favourite hobbies of 2nd graders

9. **a.** Fill in the missing numbers in this two-way table.

   b. Is there an association between the two variables?

	Boys	Girls	Total
Sports	26	21	47
Music	12	13	
Reading		20	33
Arts & Crafts	5	13	18
Video games	16		23
Cooking	2	5	7
Photography	6		
Total	80	84	164

10. The volume of a circular prism is $V = \pi r^2 h$,
    where r is the radius of the circle and h is its height.
    Solve this for r.

11. **a.** Draw a line to fit the trend in the scatter plot below.

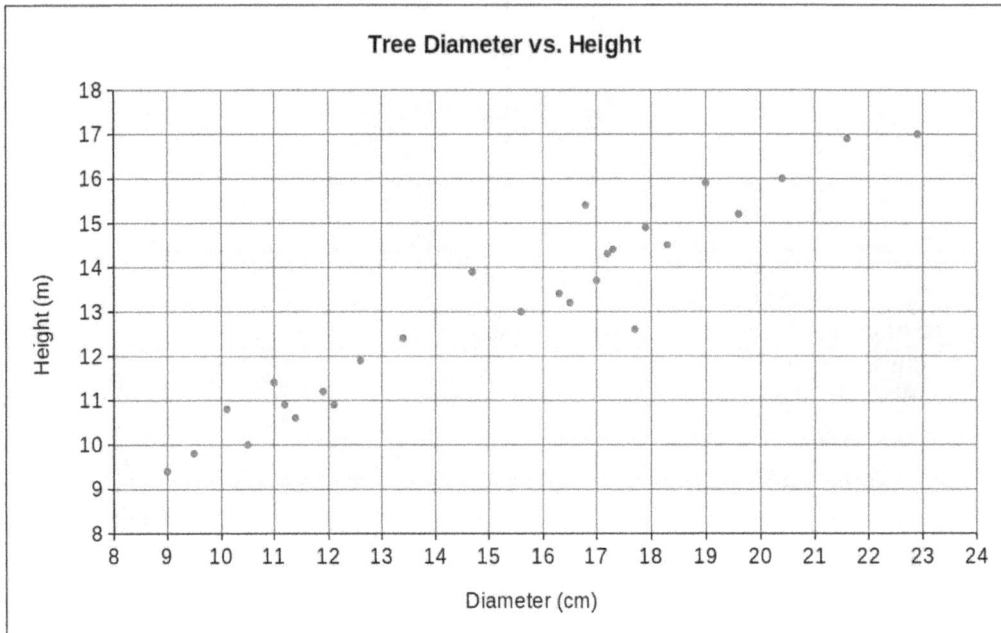

Tree Diameter vs. Height

b. Find the (approximate) equation of your line.

c. What does the slope of your equation signify in this context?

d. What does the y-intercept of your equation signify in this context?

e. Using your equation, predict the diameter of a tree that is 16.5 m tall.

www.ingramcontent.com/pod-product-compliance
Lightning Source LLC
Chambersburg PA
CBHW080252200326
41519CB00023B/6965